Climate Change Facts

By Cayden Marlow

(C) Copyright Toasted Tiger Publishing. All rights reserved.

Just a few things before I begin with the actual first chapter:

Agriculture is threatening 86% of the 28,000 species at risk of extinction, researchers said in a report launched with the UN's environment program. Low cost food is reliant on our use of fertilizer, pesticides, energy, land and water, and use of unsustainable farming methods.

"The more we drive food production, the cheaper food becomes, and the more our diets become dominated by a smaller number of crops grown intensively and at scale," Tim Benton, Chatham House's research director in emerging risks and one of the report's authors.

The way we produce food isn't only threatening the Earth's biodiversity, researchers warn. Accounting for around 30% of human-produced emissions, our food systems are also driving climate change.

Almost a third of the Earth will need to be protected by 2030 and pollution cut by half to save our remaining wildlife, as we enter the planet's sixth era of mass extinction, a United Nations agency has warned.

Time and again, scientists, experts and environmentalists have warned that the Earth has reached a crucial tipping point -- recent research from the World Wildlife Fund found the world's wildlife populations have fallen by an average of 68% in just over four decades, with human consumption behind the devastating decline.

From Amy Woodyatt, February 5, 2021, CNN.

Oh, how far we have come.

"It is mankind and his activities that are changing the environment of our planet in damaging and dangerous ways. The result is that change in the future is likely to be more fundamental and more widespread than anything we have known hitherto. Change to the sea around us, change to the atmosphere above, leading in turn to change in the world's climate, which could alter the way we live in the most fundamental way of all."
Speech to UN General Assembly, **1989**, by **Margaret Thatcher**, British Prime Minister 1979-1990

"Three in four American voters want to see the government step in to limit carbon emissions – including a majority of Republicans (55%). Voters' concerns simply aren't being adequately addressed – by the president or Congress."
Polling Memo, 2019, Frank Luntz – GOP Pollster

"…it is to me illogical to say, 'I believe in the miracles of science in terms of what it can do for our bodies at hospitals like Johns Hopkins, but then say, 'I don't believe in science when it concerns the earth."
Interview on MSNBC, 2019, Mark Sanford –Governor of South Carolina 2003-2011

Jared Diamond warns: Environmental solutions are not a "luxury" with just a cash outflow. "This one-liner puts the truth exactly backwards. ... Environmental messes cost us huge sums of money both in the short run and in the long run" and "cleaning up or preventing those messes saves us huge sums in the long run, and often in the short run as well."

The really sad thing is that some of these quotes are over 25years old, and yet, somehow, this is still not sinking in. Let me show you a few more, then you will see how spot on and correct all of these people were when they said these things. Nothing has changed since they said them, everything has gotten a lot worse instead.

"Twenty-five years ago, people could be excused for not knowing much, or doing much, about climate change. Today we have no excuse." -- Desmond Tutu

"The world will not be destroyed by those who do evil, but by those who watch them without doing anything." -- Albert Einstein

"If you really think that the environment is less important than the economy, try holding your breath while you count your money." -- Guy McPherson

"I'm often asked whether I believe in global warming. I now just reply with the question: Do you believe in gravity?" -- Neil deGrasse Tyson

"We are the first generation to feel the sting of climate change, and we are the last generation that can do something about it." -- Jay Inslee

"A nation that destroys its soils destroys itself. Forests are the lungs of our land, purifying the air and giving fresh strength to our people." -- Franklin D. Roosevelt

"We do not inherit the earth from our ancestors, we borrow it from our children." -- Native American Proverb

"We cannot burn our way to the future. We cannot pretend the danger does not exist -- or dismiss it because it affects someone else." -- Ban Ki-moon

"We must face up to an inescapable reality: the challenges of sustainability simply overwhelm the adequacy of our responses. With some honorable exceptions, our responses are too few, too little, and too late." -- Kofi A. Annan

Introduction:

Let's get you up to speed really quick, considering that this is a short read, here are a few bits that you need to know and will encounter again, later in this book:

Some species of sharks and rays could disappear from our seas altogether after a sharp drop in their numbers due to overfishing in the past 50 years. A study published in the journal Nature found that shark and ray populations fell by 71.1% from 1970 to 2018.

"Knowing that this is a global figure, the findings are stark," said Nick Dulvy, a biologist at the Simon Fraser University and a co-author of the study. "If we don't do anything, it will be too late. It's much worse than other animal populations we've been looking at." Their results revealed "an alarming, ongoing, worldwide decline in oceanic shark populations across the world's largest ecosystem over the past half-century, resulting in an unprecedented increase in the risk of extinction of these species," said the study. However, the threat of overfishing far outpaces any trade regulations or sustainable fisheries management, researchers warned.

"We can see the alarming consequences of overfishing in the ocean through the dramatic declines of some of its most iconic inhabitants," said Nathan Pacoureau, the paper's lead author.

In the report, published in the journal Frontiers in Conservation Science, the authors wrote: "Our goal is not to present a fatalist perspective, because there are many examples of successful interventions to prevent extinctions, restore ecosystems, and encourage more sustainable economic activity at both local and regional scales.
"Instead, we contend that only a realistic appreciation of the colossal challenges facing the international community might allow it to chart a less-ravaged future," the team added.

Still, they wrote, **fundamental changes to global capitalism, education and equality are required to address the problem**.

The world set a 2020 deadline to save nature but not a single target was met, UN report says:

Ehrlich is the author of "The Population Bomb," a controversial 1968 text that warned of overpopulation, predicting millions of people would starve to death.

He has since said that while many details and timings of events were wrong, the book was correct overall, telling the Guardian in 2018: "Population growth, along with over-consumption per capita, is driving civilization over the edge: billions of people are now hungry or micronutrient malnourished, and climate disruption is killing people."

As part of a bleak "prognosis," a team of 17 leading scientists cautioned that the future of the planet is "more dire and dangerous than is generally understood," and say they have conducted the assessment to clarify the seriousness of the situation the world faces.

Daniel Blumstein, professor at the Institute of the Environment and Sustainability at the University of California, Los Angeles and one of the authors of the article, told CNN it was no exaggeration to talk about a potential risk to our civilization.
"Maybe people certainly recognize it, but they don't understand the urgency, or maybe they recognize it, but they don't want to take the individual sacrifice," he said.

The time delays between ecological deterioration and its socioeconomic impacts mean people do not grasp the seriousness and timeliness of the problem, the report's authors said.

"The mainstream is having difficulty grasping the magnitude of this loss, despite the steady erosion of the fabric of human civilization," lead author professor Corey Bradshaw, of Flinders University in Australia, said in a statement.
"In fact, the scale of the threats to the biosphere and all its lifeforms is so great that it is difficult to grasp for even well-informed experts."

Wired had an article that said that by 2025 about 2/3 of our planet will experience persistent water shortages.

And none of those scientists are talking about refugees or wars, not yet. When you combine their dire outlook and add the potential for war, we may have less than 50years left.

And that is where we start the first chapter.

CHAPTER 1

In about 45 years from now, many parents will have to look their children in the eyes and have an uncomfortable and morbid conversation. This family talked about their meeting for a long time. It started when Sarah was just 6, about a year ago, she began to understand TV news and saw wars, oil spills, refugees and starving children being carried by their moms towards some desert border. She saw it, she understood, not all of it, but Sarah knew, she knew instinctively that something was terribly wrong.

Her parents did not want to have this meeting, not for a long time. They were not comfortable, as if you could get comfortable with what they were about to do. No, they were scared. Are we doing the right thing here? Is there no other option? What if we just move further north, again. What if the government has some secret plan that we could be part of? Was there nothing left we could do for our daughter? Are we really sure this is the only way?

They started slowly, talking about her favorite toy, her neighborhood friend who died three weeks ago along with her parents. It was Sarah's first funeral; she did not take it well. But it would be the last time she ever had to watch one of her friends get buried. It was not the best moment to hold this family meeting, or, maybe, because of that funeral, this meeting had to happen now. These were very confusing times.

There was no other choice, this was the day. Sarah had been prepared by her parents for this important meeting and talk. They told her that next weekend, next Sunday, we all get together, we all need to talk about our future. Sarah heard it several days in a row, so she was a bit excited to have her first adult and grown-up talk with her parents. Mom even prepared her favorite juice for the occasion, that weird colored powder that you put in water, it makes it all look bubbly and green, Sarah liked drinking it. She even dressed up in her beloved cartoon character skirt and shoes. Sarah had her favorite doll in her hand, the one with the curly brown hair, and placed her in her lap.

After a low voice opening from her dad, and some tears in her mom's eyes, Sarah was no longer excited. She was more scared and confused than she had ever been before. She never saw so much fear in her father's eyes.... but she brought up the courage as she asked him:

"But, daddy, why do I have to die?"
"Well, sweetie, mom and I didn't really believe it would come to this..."
"How come?"
"What do you mean?"
"How come you didn't want to believe it?"
"Well, I guess, we just didn't listen..."
"Like I sometimes don't listen?"
"Yeah, honey, something like that..."
"I see, how much longer do I have?"
"Well, it looks like we are all going to end real soon."
"But why daddy, why do we all have to die?"
"We were greedy, we all wanted more money, none of us wanted to change our system...."
"What's 'system', dad?"
"That's a way we all agreed we would do things, you know, it's like how we set the dinner table?"
"Oh, where I sit in between you and mommy?"
"Yeah, like that, see? We have a system at our dinner table. Same idea. So, we have a system on how we create and count money, and then, well, I guess we ran out."
"You ran out of money?"
"No, sweety, not us, but yeah, kinda like that. The whole country, the whole world, ran out of money, so we didn't have any money left over to save our planet."
"Can you use my piggy bank?"
"No hun, that won't help us now. It is too late."
"Too late?"
"Yes, the heat is so hot, the ice melted everywhere, we hardly grow any food, and our drinking water is full of toxins... it's the end, we can't reverse the damage we caused...."
"Why didn't you stop it daddy?"
"I don't know baby; I really don't know."
"You said it was money?"
"Yes, it was."
"We don't have enough?"
"No, we ran out."
"But who makes the money dad?"
"We do sweety, humans created money..."
"Why don't you just make more of it then?"
"We tried, but we couldn't agree on how to do it."
"Why not daddy? Why can't adults agree on it?"

The mom hugged Sarah really tight, with tears in her eyes and they all started a big family hug. Sarah squeezed her doll even harder. The father was crying, mom wiped tears off her daughter's cheek and then slightly nodded to her husband, before giving him one more kiss on his mouth. The dad got up and walked to the closet, pulled out a little case and opened it. He took out his gun, loaded it, took off the safety and cocked it. He turned around and walked towards his family and with tears falling from his eyes, as he looked into Sarah's stunned face, he looked at his daughter's beautiful eyes, then he shot her point blank into her forehead.

The violent impact threw the little girl several feet backwards, her head was wide open, her face and dress were covered with blood, so was her doll, now laying several feet away from her. Her mom screamed, placing her hands in front of her mouth, looking up at her husband. He already pointed the weapon at his wife's head and pulled the trigger. Another loud boom thundered through their small living room. He looked at his dead daughter, then at his dead wife, all the blood on the floor, placed the gun in his mouth, and pulled the trigger one last time.

At this moment millions of people had already died because of global warming and all its effects, from toxic lake and tab water to mudslides to Pacific islands and countries sinking and wars over crops, etc...

But for this family it was much easier this way. After all, just in their neighborhood street alone, south of Austin, Texas, just last week the coroners came three times, pulling over 13 people from their homes, all dead by suicide. This was just one neighborhood; it was happening all over the south, from flooded Florida, scorched Arizona to the Carolinas. News from last month said that over 2000 people in drought ridden Australia all killed themselves, on the same day.

A mass suicide that wiped out a small mining town that nobody ever heard of before. Even the preacher of their small make-shift church had his wife and son drink the poisonous concoction the towns people brew the day before. There was no hope left, no clean water, no food, just dusty winds howling through the streets.

Who wants to truly wait for the last famine on earth before you accept your fate and watch your children starve to death or puke their bloody lungs out from all that toxic water and the soot filled air that grew the tumors in your daughter's lungs and stomach?

Many home-made recipes and formulas will be circulating through communities, on how to best kill yourself, your family, your kids, and pets, in their sleep, without pain. Families get together to discuss the best way to prepare prescription drugs and how to mix them with alcohol, and some cool aid, to help their kids drink it.

Yes, that's the future. There is no doubt about it. There is no real discussion about it. There is no way of stopping that from happening.

We may be discussing the Paris Climate Accord, but that is not going to stop the human catastrophe we created here, far from it.

Let me show you,

CHAPTER 2

The abysmal catastrophic failure of cosmic proportions that destroyed entire habitats, ecosystems and species is unfathomable – we're killing everything around us and we're too stupid to save ourselves.

We should have stopped all of this about 30 years ago. Yeah, I know, some really smart people and scientist did warn us, 30 years ago, but we didn't listen. Want to try again? Want to listen now?

WHY is it so hard not to pollute the planet? Well, honestly, it's a LOT cheaper to pollute than not to pollute. When the industrial revolution started, nobody was thinking about global warming, poison in the river, oil in our aquifers or anything even remotely related to our environment. We just dumped everything we didn't like or needed or knew what to do with in our own back yard. Nobody cared that all that sludge and run-off is eventually going to poison our rivers, which will poison the fish, which then die, so the people that lived off the fish are now starving, sick, ravaged by cancer, have no health insurance and some new big Chemical or Oil company is paying some politician to allow them to dump even more poison in the same river. Nobody cared back then. Profit was all that mattered. It was all about money. And it still is.

At this time, we are so far into our industrial love with money and profits that it's difficult to imagine that you had to rebuild a factory, just because some pipes let thousands of gallons of industrial sludge run straight into a river, and some smokestack polluted the nearby towns. Installing filters and new pipes? What are you crazy?! You have any idea what that will cost? And so we let it slide, we kept on going, who cared, the planet was big, the rivers would never run out of water, crops would never fail and the ocean was so big, who cared about some extra million gallons of toxins, no problem, the ocean can handle it....

Sometime in the mid-70s the mighty military industrial complex was too rich and so influential that it became ever more difficult for politicians to ignore big oil or other big industries. They had so much money that they could afford to bribe anyone. And since that was a bit frowned upon, they quickly changed tactics and began to call it *lobbying*. Same thing, different name.

Where is all of that greed leading us to? Well, here is the first insight that should shock all of us:

Eight million metric tons. That's how much plastic we dump into the oceans **each year**. That's about **17.6 billion pounds** — or the equivalent of nearly 57,000 blue whales — every single year. (conservation.org)

Or look here: https://www.visualcapitalist.com/all-the-worlds-metals-and-minerals-in-one-visualization/

The global economy's appetite for materials has quadrupled since 1970, faster than the population, which only doubled. On average, each human uses more than **13 metric tons of materials per year**. (more than a ton per month)

In 2017, it's estimated that humans consumed **100.6Billion** metric tons of material in total. Half of the total comprises sand, clay, gravel, and cement used for building, along with the other minerals mined to produce fertilizer. Coal, oil, and gas make up 15% of the total, while metal makes up 10%. The final quarter are plants and trees used for food and fuel.

More material per person, means more boats to ship it all, means more bunker fuel for those boats, means more pollution, means faster warming, means faster sea level rise, means sooner and more crop failures, but we are getting ahead of ourselves.

Nobody saw all that coming in the 1970s, well, there was this one study....

We all loved it. Those of us who grew up in that era remember ever growing economic mobility, happy parents, low cost college tuition and green parks everywhere. Then came Chernobyl and some people began to wake up. It was 1986 and suddenly the world had become a lot smaller. Maybe, just maybe nuclear power and reactors were not quite as safe as we all thought, or as our governments told us they were. We had been to the moon, we conquered space! surely, we could build indestructible nuclear power plants, couldn't we? Now we all started thinking: What happened somewhere else could quickly come home to haunt us all. Really?

Nah, bullshit, they just didn't built it right, it was doomed to happen, don't worry about it, keep going, we're not that dumb, we build better ones, who cares, it's far away, yeah, keep going, pollution is not that bad....look, there's plenty of fish in the ocean, no problem.. why change? Nuclear power, ah, nothing like that will ever happen again, never. We saw it, we learned, we're gonna do better next time, trust me.

It's not like we're gonna build nuclear powerplants right by the ocean, no way, they could get hit with a tsunami or something, no way, that would be crazy, we will never do that.... right?

What kind of toxic pollution grave have we created?

CHAPTER 3

Where are we today? Even if we stopped *everything*, all of it, we stop all factories, stop all cars, stop all planes and boats, turn off all electricity production, all cows stop farting,... as if mankind disappeared from earth overnight, we are gone, turned off everything, poof, lights out everywhere. Yes, I mean all of it. Even if we could make that happen....

Well, even if we did that, our planet would still continue to get warmer for the next 100years or more, ocean/sea levels will still keep rising and swallow Miami, London, Bangladesh etc...

IT WILL NOT STOP WARMING!

We put ourselves in a position that seems impossible to get out of. We are so far into polluting our planet that it is not possible to stop this warming trend. You get that? It's NOT POSSIBLE, for a long time to come.

Doesn't that mean that our politicians talking about a gradual reduction in pollution are really not helping the issue at all? Yes, yes, my friend, that is exactly what that means. Any actions to *gradually* decrease some minor parts of global pollution are futile. Most of the rhetoric coming from Washington and the UN and EU leaders etc. is utter nonsense and totally useless.

Let me show you how bad it is already and why it will get warmer, even if we were not here anymore. We already torched our planet.

Warming climate crisis conditions thaw permafrost, which is causing the tundra to explode into the blue sky. There are dozens of pictures from giant holes, the size of football fields, hundreds of feet deep, where the methane gas exploded underground. It leaves these giant holes all over the tundra. Just google: tundra methane explosion.

The methane that was trapped under a frozen layer of dirt, got out and blew up, because the frozen layer of earth above it, well, that's NOT frozen anymore. The so-called "Perma"-frost is no longer *permanently* frozen. This problem is also all over Alaska, where oil pipelines are built and anchored onto frozen ground, and, yes, even cities are built on that *'frozen'* ground. Guess what, that ground is no longer permanently frozen. It is melting, shifting, opening up, exposing methane gases, creating sink holes....and sooner or later an Alaskan pipeline is going to burst because of it. And this is just one picture, these holes are all over the tundra, more and more of them. Just google it.

Now, even if we were to turn everything off, it will still get warmer, because there is more and more methane gas escaping and getting trapped in our atmosphere, which in turn heats up our planet. *In the first two decades after its release, methane is **84 times more potent** than carbon dioxide. While methane doesn't linger as long in the atmosphere as carbon dioxide, it is initially far more devastating to the climate because of how effectively it absorbs heat.* (edf.org)

We are at a point where it is too warm already to stop the melting of the permafrost, ergo, more gases will come out and create more heat, which will warm up the planet, and then melt more permafrost and melt more ice sheets, melt it all, and then the heat will really get hot because by that time there won't be any ice sheets left to reflect the sunlight and our planet will absorb all of the sun's energy. Oceans will continue to get warmer because of that, which will increase the number and strength of hurricanes and typhoons, which will devastate coastal cities, which are getting flooded anyway because of sea level rise. The vicious cycle began 20-30 years ago, and we are wayyyy too late to the party.

Alaska is on fire, Brazilian Wetlands (get it, they are wet!!) they are on FIRE, California is on FIRE, the arctic circle has forest fires!! How on earth is ice burning? I did say arctic circle. Arctic means cold and ice and snow, right? Well, apparently not anymore.

Hundreds of wilfires ignited, smog has prompted several regions to declare states of emergency and smoke has blown across major cities like Novosibirsk, blotting out the sun and making it difficult for some people to breathe. This soot can be harmful to humans and animals, entering the lungs and bloodstream.

The arctic is on FIRE. This is the point where you should scream your head off.

Try to wrap your head around that. For over a few thousand years mankind has lived in these areas and never knew anything but ice and snow. Now, now it is all melting away, greenery, bushes, brush and trees come to live, then they dry out, it's too warm, then forest fires start with just a bit of lightning and it all burns and increases greenhouse gases and makes our planet even warmer than it already was, which leads to more ice melting and warmer oceans and .. and... and....

Then you get this:

630 square miles, the **largest** chunk of ice in more than 50 years has broken off the Amery Ice Shelf in **Antarctica**. The new iceberg, which broke off in September 2019, weighs **315 billion**

tons. Officially named D-28, the iceberg is more than 630 square miles in area and nearly 690 feet thick. (cbsnews.com)

A piece the size of Manhattan or Los Angeles, was not part of our scientist's predictions.

All the prediction models that talk about the next 50 years or the next 100 years are wrong. None of them take the compounding effect into account. What is that? Well, that's like making dominoes fall, but with a lot more umph at the end. One thing leads to another, making that next thing worse, making it happen faster, which leads to the next thing happening a lot sooner, which makes it all worse... Greenland Ice sheets are sliding off into the ocean... because they are melting _underneath_. Yes, UNDER the ice, where it is frozen onto the rock, that part is melting, because warm ocean water is getting underneath it. And from the top, the ice is melting, which leads to little puddles, those become ponds, that become lakes, that let warm water sink into the ice, heated by the sun.

*A new study published in the journal Nature provides some answers, and it is not good news: The rate of melting we're seeing today already threatens to **exceed anything** Greenland has experienced in the last 12,000 years.* (nature.com)

Right now, Briner says, current melt rates track closely with this **worst-case scenario**. What happens to Greenland's ice sheet and others around the globe will determine what the future holds for the millions of people living along the world's coasts.

In terms of its potential to raise sea levels, Greenland is the world's second most important ice sheet, behind only Antarctica. **Greenland's ice sheet contains enough water to raise global sea levels by 24 feet**. Over the last 26 years, melt water from Greenland has raised sea levels by 0.4 inches, and it is currently the world's biggest contributor to sea level rise.

Antarctica holds enough water to raise sea levels by about 200 feet, and scientists have recently discovered **alarming vulnerabilities** in some of its most important glaciers.

Now that warm water is sinking *into* the ice in Greenland, literally cutting a hole, ever more, ever faster, going ever deeper into the ice, until it reaches the rock underneath. Now the ice shelves are melting from the bottom and lose their grip on the rock below, they start to move a little, they start to crack, they start to slide, which makes them fall into the ocean a lot sooner, which breaks all prediction models! Holy tossed salad, ice shelves are melting from the bottom up?! What climate model predicted that?

What does mean? It means that even if we all disappeared over night, today, and turned off all our lights and factories and stopped all greenhouse gas emissions.... it would still get warmer, it would still get worse, sea levels are still rising, ice sheets are still melting and still more gases are released into our atmosphere, even if it isn't us doing it. We did it all already, a long time ago, but now it doesn't even need us to add to the problem anymore. We created such a huge catastrophe that it is impossible to stop.

How warm will it get? How bad will it get? When will it get that bad? The when is easy to answer, the when is NOW! How bad will it get? It will be worse than WWI, and WWII and Covid-19 combined, and it will be far worse than your worst estimates can currently comprehend. Here is why, a heatmap from Impactlab.org shows temperatures of over 90degrees, for about 6months of the year, all over the south of the USA. That is Texas all the way to North Carolina. For six months out of the year, it will never cool down below 90degrees. That's petty scorching hot they say around 2080, but we already know that most models are far too generous with their time frames. Who wants to move to Texas? Or Arizona, New Mexico? Are you going to turn on your air condition, to use up more energy, more coal fired power plants need to spew out more toxic gases so you can have electricity to cool your 110 degree 95% humidity living room in the mid-west and complain that a bottle of water costs $158.50 at that time? Really? Is that where we want to go?

And does anyone really believe that New Orleans will still be there in 30 years from now? Really? When you look at that map you can see migration from the south to the upper Midwest and from the lower west to the inner states. Millions of people will have to leave their cities behind. Louisiana is a thing of distant memory in about 30 years from now. Mardi Gras? What is that mommy? Here, let me show you some old pictures from the internet... We refer to some places as the "low lying areas" ...... How insane is that? Just tell the truth. Call them: Soon to be gone areas.

You should look at **350.org**. The name comes form the idea that our atmosphere can handle about 350 parts per million of CO_2 in our air. Anything below that is good, it won't warm our climate, anything above that number and our climate will keep warming. Too many CO_2 particles in the air warm up our planet. Yeah, I know, science and math, stop complaining and keep reading. It's easy:

350 is good, less than 350 is better, and anything above 350 is really, really incredibly bad... got it? Good. Now guess what, surprise......

By 2007 we reached over **385** parts per million. That was 13 years ago. We are in deep doodoo, up to our necks. The organization 350.org put out a great documentary, they went to Washington, they met with politicians, they talked, blah blah blah blah, and, of course, nothing changed.... The latest numbers that came out of our atmosphere are:

*In **2019**, CO2 concentrations crossed **415** ppm in the atmosphere for the first time ever!... even if we stopped emissions now, many glaciers will still keep melting....*

Can you see it? Even if we stopped now, if we stopped ALL emissions, we STOP *EVERYTHING*, and it will still get warmer, and warmer, and warmer.

On a serious note though, don't be too alarmed about all of this, I don't want you to despair yet. You really don't have to worry about all of the stuff above. It's not so bad actually. Keep reading, it's gonna get a LOT worse...

CHAPTER 4

GLOBAL WARMING FREEDOM FIGHTERS.

The ARAB SPRING started because in the region, and especially in Egypt, they had massive food price increases in 2010, because Russia took their wheat off the world market, because a drought reduced crop yields on a global scale, that led to price increases, now bread was too expensive in Egypt, and they love their bread... so they started to demonstrate, that led to riots, that led to people dying in the streets, that almost toppled an entire government. See?

Those 'oppressive regimes' did not have the ability to handle the drought, crop failures, lack of water, with that the lack of hope, and they certainly didn't expect global warming to cause Russia, usually a large exporter of wheat, to abruptly stop all their exports. And if your government can't provide stable prices or make sure that you don't have any food shortages, well, then they are considered 'oppressive', right? Let's riot!

How often do I have to hit most people over their head with a big iron frying pan before they even begin to understand the term: GLOBAL, that part means that the effects from all of our pollution will be felt, you might have guessed it, on an earthly global scale! That means everywhere, all over the planet, all at the same time.

We are all connected on this planet. It is happening on a GLOBAL SCALE.

Yes, if you are in America, and you thought global warming wasn't affecting you, then you just don't deserve this place. A kid joined the military, US Marines, ended up in the middle east, had to fight ISIS, died in a deserted village, all because we voted for politicians that don't understand money and didn't care about Global Warming and could not see themselves helping our fellow humans to avoid a drought, that created food shortages, so farms failed, a **million people** moved from those farms and areas to Damascus and Aleppo, looking for jobs, looking for food etc... and over 100thousand young men ended up joining terrorist organizations that promised them food, water and bread, and we called them ISIS. Those are the starving farmers' sons that our kid had to go and try to kill. Do you see it now? We are all connected. You can't pretend that we are not connected.

A drought in one place of our planet will have an effect on many other places on our planet. How many of those dominoes do we want to add to this line of death??

Please watch: "*The Age of Consequences*", available also on Amazon, Prime Video.

Russia could not grow enough wheat to sell on the global market, due to a drought. That caused the Arab Spring. Another drought brought famine to the middle east that lead to ISIS being able to recruit thousands of young men as fighters. They just wanted to have food and water, and a job. You see? Rain. Water. It leads to refugees, it leads to economic collapse, countries running out of money, unable to feed their citizens. It all leads to war. Global Warming leads to war.

There is this small "low lying" land called Bangladesh. It kinda sits between India and China, who border Pakistan. India knows that global warming will destroy Bangladesh. Completely. India was really smart, they built a wall, the kind of wall that makes most governments green with envy, sharp razor wire, multiple layers of steel beams sticking out, armed guards, watch towers, drones flying with AI to observe it all, and they regularly shoot and kill about 80 people a year trying to cross into India. Yep, that is a Global Warming Wall.

That's not so bad, right? Well, not YET it isn't it. But very soon the sea level rise will reach further and further inland, and then about 30 million global warming refugees from Bangladesh will have to try and escape from drowning and starving. Exactly where do you think those people will go? Where would you like them to go?

Did you build a city for them already? Did anyone built a city for 30 million people that we know are coming? Does India really believe they can keep out all those people? Or China? What if they can't agree on how to handle all that? Yeah, India, Pakistan and China, they all have nuclear weapons. Ooops. Can global warming lead to nuclear war? Yes, indeed, yes, it can.

And when the water levels swallow Bangladesh, they also swallow Louisiana and New York.

Just who came up with the idea that mankind had nothing to do with global warming? Who was it that concocted this bizarre idea about a solar cycle? Can you guess? Really, no idea? How about we talk about the real cause of global warming, then you might see the connections. Ready? Good, here is the main reason for global warming:

MONEY

A colorful cast of characters has made a living out of denying the science of climate change. These so-called "experts" often start out their statements with "I'm not a climate scientist, but..." before launching into a series of carefully rehearsed talking points meant to confuse the public on the climate change issue. Many of them are well-paid operatives of organizations like The Heartland Institute, CFACT, and Americans for Prosperity, which take contributions from fossil fuel corporations —
.....including ExxonMobil, the Koch Brothers and their company Koch Industries — who seek to delay or block any substantial government policy initiatives meant to curb fossil fuel emissions or hasten the rapid growth of cheaper, cleaner sources of energy like wind and solar power. (BeforeTheFlood.com)

*Solar activity is **definitely NOT the cause of climate change**. The activity of the sun has been accurately measured using both ground and space-based systems for <u>decades</u>, and no significant change in solar output has been detected that could be responsible for the dramatic warming now occurring on the Earth.*

Several other indicators confirm this observation. First, the planet is warming more, not less, during the **winter** and it is also warming faster at the **poles**, not at the equator — the opposite of what would be expected if solar activity was driving global warming. Most important, as shown in the Earth's temperature readings, the top of the stratosphere (the upper level of the atmosphere) is cooling, while the lower atmosphere is warming. If the sun were the culprit, we would see hot spots of heat on the top of the atmosphere... (BeforeTheFlood.com)

Here is a list of studies if you are truly interested:

https://www.carbonbrief.org/the-most-influential-climate-change-papers-of-all-time

Let's look at more connections.

CHAPTER 5

Pollution, it all dissipates, doesn't it? Let's get away from global warming for a moment and just talk about pollution. Yes, CO2 is a toxin, it pollutes our air, never mind that it also traps heat in our atmosphere. Let's look at pollution without the warming effect, just a little bit.

Hang in there, this is not boring, this is really short, just a few bits of info you should know about before we go on.

The U.S. Energy Information Administration estimates that in 2017, the United States emitted 5.1 billion metric tons of energy-related carbon dioxide, while the global emissions of energy-related carbon dioxide totaled 32.5 billion metric tons. (Usgs.gov)

Globally, **91%** of the world population is exposed to unhealthy levels of pollution. (datatopics.worldbank.org)

Air pollution is a major global environmental risk to our health and food security. It is estimated to cause about 3.7 million premature deaths worldwide and destroys enough crops to feed millions of people every year. (ral.ucar.edu). You see? We are already killing ourselves.

The World Health Organization, WHO, estimates that around 7 million people die every year from exposure to polluted air. Ambient air pollution alone caused some 4.2 million deaths in 2016, while household air pollution from cooking with polluting fuels and technologies caused an estimated 3.8 million deaths in the same period. (who.int)

The groups most affected by air pollution are people of color, elderly residents, children with asthma, and people living in poverty. Vulnerable populations may experience more health effects because these populations already have higher rates of heart and lung conditions. (pca.sate.mn.us)

After years of improvement, US air quality has gotten **worse** since 2016, report suggests. A new report by a nonprofit research group found air quality in the United States has worsened since 2016, even as officials in the Trump administration have claimed "the environment is getting cleaner". Nov 1, 2019, (usatoday.com)

Put a company name next to the pollution, so you can see the money behind it:
    Saudi Aramco, Saudi Arabia.
    Chevron, US.
    Gazprom, Russia.

ExxonMobil, US.
National Iranian Oil Co., Iran.
BP, UK.
Royal Dutch Shell, Netherlands.
Coal India, India.

These are the biggest polluters on our planet. Can you smell the money?

However, this one little part, the one that no hoaxer, skeptic or denier can find a way around when they try to spew their ignorance, that minor part is this:

By drilling for ice cores and analyzing the air bubbles, scientists have found that, at no point during at least the past 800,000 years have atmospheric CO_2 levels been as high as they are now. That means that **in the entire history of human civilization, CO_2 levels have never been this high**. (climatecentral.org)

Let me try to emphasize that just a tiny bit for you to recognize the implications, you know, just in case you come across somebody who has been corrupted by the oil, gas and coal industry to try and tell you that our pollution is perfectly normal for our planet and blah, blah, blah....

It says: **NEVER** before, since the beginning of mankind, never before, have CO_2 levels been this high, EVER, on our planet. Now that is just the accumulative effect of everything that we have done since the beginning of the industrial revolution.

That is not referring to the last 150 years. What if we are just talking about since the 90s? Here you go:

We've emitted more CO_2 in the past 30 years than in all of history.

(thecorrespondent.com)

UN Scientists See Largest CO_2 Increase in 30 Years - (ourworld.unu.edu)

More than half of all CO_2 emissions since 1751 emitted in the last 30 years. (ieep.eu/news)

"Already in 1990, we knew about humanity's impact on climate change, and yet we've emitted more than half of all emissions since 1751 in the last 30 years alone. If we had taken action then, today's young people and future generations would face a much easier challenge. We cannot afford another decade of inaction and have to flatten the emissions curve......"

And again, ignoramus sapiens will say:
"Ah, you don't really know that."
"They could have read it wrong"

"You can't just look at snow and figure these things out, that's not how it works"

"Well, I don't really care, there is that one lady, some scientist, I think, and she once said that we are not the cause, so, I believe her, and I ain't changing nothing"

"They are just scientist, what do they know?"

"They all got some kind of agenda, I ain't listening to that stuff."

I am just trying to make you understand that it is not a very good idea to dump hundreds of millions of tons of toxins into our environment and to expect nothing will change. That is insane. Of course it has an impact. How much of an impact is irrelevant, just stop polluting. Why is that part so difficult for most people to grasp? Well, they somehow think that all this pollution is helping them to keep their jobs and house and whatnot. As long as those things don't change, they don't want to look at anything.

NASA says:

"The composition of Earth's atmosphere has most certainly been altered. Half of the increase in atmospheric carbon dioxide concentrations in the last 300 years has occurred since 1980, and one quarter of it since 2000. Methane concentrations have increased 2.5 times since the start of the Industrial Age, with almost all of that occurring since 1980. Changes are coming faster, and they're becoming more significant." (climate.nasa.gov)

In little more than a century of frenzied pollution dumping, humans have altered our planet's atmosphere at a rate dozens of times faster than natural climate change. Carbon dioxide is now more than 100 ppm higher than any direct measurements from Antarctic ice cores over the past 800,000 years, and probably significantly higher than anything the planet has experienced for at least 15 million years. That includes eras when Earth was largely ice-free.

Not only are carbon dioxide levels rising each year, they are accelerating. Carbon dioxide is climbing at twice the pace it was 50 years ago. Even the increases are increasing. (grist.org)

There is that compounding cascading effect I told you about.

How much do you know about boats?

CHAPTER 6

How can BOATS be worse than all the cars in the world?

Consider this, what if I told you that there are well over 60.000 container ships on our oceans at any time, with over 400.000 crew members/sailors, and that doesn't even include the thousands of oil tankers?

The top 10, that just the number, ten, ten boats, the 10 biggest boats on our oceans spew out more toxins than 50million cars. Yep, those numbers are correct, read them again. OK, I lied, the numbers are worse:
"It has been estimated that just **one** of these container ships, the length of around six football pitches, can produce the same amount of pollution as 50 million cars. **The emissions from 15 of these mega-ships match those from all the cars in the world.** And if the shipping industry were a country, it would be ranked between Germany and Japan as the sixth-largest contributor to CO2 emissions."

It's difficult to imagine how much carbon dioxide that is, especially because ships operate so far out at sea. Alongside CO2, ships also release nitrous oxides (NOx) and sulfur oxides (SOx), which are highly toxic chemicals that are proven causes of acid rain. While manufacturers are slowly designing less environmentally detrimental cars, the oceanic freight industry is subject to fewer regulations.

(https://inews.co.uk/news/long-reads/cargo-container-shipping-carbon-pollution-114721)

Are you getting the idea? 15 boats are worse than all the cars in the world. That is insane.

There are some horrific photos on the internet of containers that fell off their cargo ships. Entire boats listing to one side, containers clinging on to their last hope with just one screw left, containers floating all over the ocean. If you google it, there are estimates that anywhere from 2000 to 8000 containers are LOST at sea, every year. Every. Year. Sinking, after a while, to the bottom of the ocean. Shipping lanes across our planet are littered with rusty, rotting, molded and toxic crap shipping containers.

The boats, or better yet, their engines, are a marvel of engineering magic. They are so good, that you can drink a bottle of vodka, pee in the tank of those engines, and they will run and hum as if they loved it. Because they do. Boats of this size run on bunker fuel. You probably never heard of it, but you have seen it. Big black plumes of smoke, that look as if the boat was on fire. It is called 'bunker' fuel because in the old days of steam ships they held their coal in 'bunkers' on the boats. Now, well, nowadays not much has changed, why should it?

Most ship engines have been designed for top speeds ranging between 20 and 25 knots per hour, which is between 23 and 28 miles per hour. A Panamax container ship can consume **63,000 gallons** of marine fuel <u>per day</u> at that speed. Freigtwaves.com

Oil tankers, container ships, and cruise liners guzzle massive amounts of fuel. This "bunker" fuel is chock-full of sulfur, which means it emits more poisonous gases and harmful particles when burned than does motor-vehicle fuel.

Remember, we are just talking about toxins, not global warming.

On top of the pollution caused by their exhaust fumes, cruise ships have been caught discarding trash, fuel, and **sewage** directly into the ocean. Apr 26, 2019, forbes.com

Yes, that means when a cruise ship that burns this horrible toxic laden bunker fuel, spews out more CO2 and sulfur than 200.000 cars, loaded with 3000 passengers....every time these passengers use the toilet, all that goes straight into the ocean. Lovely, isn't it? They should call it a poop cruise.

U.S. law allows cruise ships to dump raw sewage in the ocean once a ship is more than three miles off U.S. shores. Ships can dump treated sewage anywhere in the ocean except in Alaskan waters, where companies must comply with higher state standards. (eu.oceana.org)

WHY on earth would you allow that?! Are you disgusted yet? Good, keep reading.

An average-sized cruise ship with 3,000 passengers and crew produces 30,000 gallons of sewage every day and 255,000 gallons of dirty water from showers, sinks, laundries, and dishwashers.

Carnival Corporation, the world's largest luxury cruise operator, emitted nearly 10 times more sulfur oxide (SOX) around European coasts than did **all 260 million European cars in 2017**, a new analysis by sustainable transport group Transport & Environment reveals.[1] Royal Caribbean Cruises, the world's

second largest, is second, **yet four times worse** than the European car fleet.
   (transportenvironment.org)

   That means that the 47 ships from Carnival Cruises, emit more toxins than 200 million cars!
   When are they going to pay for all of that damage??
   https://www.geekyexplorer.com/cruise-ship-pollution/
   Yeah, let's book a cruise, shall we? Sure, as long as YOU pay for the costs to clean up all the pollution that your trip has caused. As long as the cruise ship company pays for the clean up costs. In other words, your cruise should not cost $600, it should cost $27.000. How much longer are we willing to let polluters get away with killing us all? When you willfully and intentionally pollute OUR environment, our planet, for the sake of and with the argument that you have to increase shareholder values, stock options and bonuses, then you are running a criminal enterprise, aren't you?

   And you thought some old black fume diesel truck on the freeway was an eye sore? Just google it: "bunker fuel boat pollution". You will see some really awful and disgusting pictures of thick black smoke coming out of the smokestacks from these boats. Imagine over 60.000 boats like that. Actually, you can follow them, you can see them all, right here:
   https://www.radar-tracker.com/marine_traffic/

   Please, I want to encourage you to watch the movie. **Freightened.com**. Not only is that movie very well done, it is fascinating what you can find out about our shipping issues.

   They want to reduce vehicle emissions by a few percentage points? Really? But we also add new boats to our worldwide shipping fleets every year. And remember, 10 big boats can spew out more poison than a few hundred million cars.......the math doesn't work.
   But it's too late, we are warming our planet ever more, and the next question is how do you control and enforce these new limitations? And the same principal applies as before. If you reduce the output per boat but you increase the number of boats, you have no effect on reducing the pollution they spew out as a whole. Their numbers and math make no difference and will just continue to kill us all. And none of that decreases all the pollution that we already spewed out into our atmosphere. We already have too much pollution.

The numbers are staggering: There are 5.25 trillion pieces of plastic debris in the ocean. Of that mass, 269,000 tons float on the surface, while some four billion plastic microfibers per square kilometer litter the deep sea. Scientists call these statistics the "wow factor" of ocean trash. (nationalgeographic.com)

Remember, even if we stopped all these boats, all of them, right now, our planet will continue to get hotter and hotter, because of the pollution and toxins we already spewed into our atmosphere. All the crap that these boats put into our atmosphere is already there. Slowing down the boats is going to reduce their *additional* and future emissions a bit, yes, but it will still get warmer and we are still building more boats.

It's like going to a wooden outhouse shack to poop. You open the door, and crap is piled up to the ceiling, falling out the front......and your friend says: "It's ok, don't worry about it, just poop a little less than usual...." That idea stinks.

We have more to talk about.

CHAPTER 7

Industrial Pollution, nah, that can't be that bad....

Let's talk about coal, just for fun. When coal is burned it releases a number of airborne toxins and pollutants. They include mercury, lead, sulfur dioxide, nitrogen oxides, particulates, and various other heavy metals. The burning of fossil fuels releases greenhouse gases into the atmosphere, increasing levels of CO2 and other gasses, trapping heat, and contributing to global warming. Coal-fired power plants release more greenhouse gases per unit of energy produced than any other electricity source.... (greenamerica.org)

The burning of coal is responsible for **46%** of carbon dioxide emissions worldwide and accounts for 72% of total greenhouse gas (GHG) emissions from the electricity sector. (endcoal.org)
The main problem of the coal mining process is the pollution of the nearby waterways by that which is brought to the surface from a coal mine alongside with the extraction of coal. ... After coal is burned, coal ash is left behind spreading toxic substances such as arsenic, mercury and lead in the environment. Jan 22, 2019
(environmentalpollutioncenters.org)
The only reason coal was always cheap, is the fact that nobody adds the costs for cleanup, dead rivers, health care and a human death toll due to coal. Our economic system and accounting methods for environmental damage and cleanup is totally flawed. If you add all the death, damage and destruction of coal, oil, and gas to their base unit price of energy, ALL alternative and sustainable energy ideas would be far cheaper than their polluting counterparts! It's like buying a pack of cigarettes and each time they ask you to add another $90.00 for your next x-ray.
But instead, we pay high taxes, so our government can give subsidies to oil and gas and coal companies, to prop them up, and then we have to pay even more money for our health care costs, but we think we're getting cheap energy, and this is the way it should be.

Powerplants are really disgusting......if you ever see one, or google it, you can see those thick plumes of grey death escaping at a pretty good clip when they come out. And they come out all the time, it never stops, they just keep pumping toxins in our air, unfiltered, unapologetic and without remorse. Not one executive working for those companies would allow their kids to breathe the crap that comes out of their smokestacks.

I haven't seen a coal mine CEO or powerplant CEO go to jail yet for polluting our environment. Why not? Seriously, why not?

As long as there are no real consequences for the operators of these companies, then there is no reason for them to stop polluting. If all it ever comes down to is money to pay a fine, then money is all we get, but they will keep polluting. Polluting is cheap and doesn't pose a threat to the livelihood of the company executives. Our laws are for kindergarten cops.

1.2 trillion gallons of untreated industrial waste is pumped directly back into the U.S. water supply every year. (https://brandongaille.com/8-industrial-pollution-statistics/). Yes, read that number again, it says <u>TRILLION</u>.

Americans only account for about 5% of the world's total population. America accounts for **30%** of the world's waste through industrial pollution while consuming 25% of the world's resources. Almost **50%** of all the American lakes are affected by industrial pollution to the point where they are unsafe to use for swimming, fishing, or other forms of recreation.

Remember the part about Bangladesh sinking due to sea-levels rising, and all the refugees that will be coming...? India has the fifth largest electricity generation sector in the world, of which two-thirds comes from coal. Emission standards for power plants in India lag far behind those of China, Australia, the EU and the USA. For key pollutants like SO2, NOx and mercury, there are *no prescribed emission standards* in India.

Despite the Indian government's warnings, most Indian coal fired powerplants had **not** installed any kind of pollution reducing technologies by mid-2019. (iisd.org)

What about landfills? You know, the place we call "AWAY". Yes, that's we all call it.

"Oh, sweetie, that is rotten, don't eat that, just throw it away"
"Throw it out, that stuff smells"
"We just threw it away after we were done with it"

Out, out and away. There is no such place as *away*. But it does sound very clean and convenient.

According to the Environmental Protection Agency (EPA). Due to land pollution, the Earth loses approximately 25 billion tons of valuable topsoil each year. Sep 4, 2012 (poopy.org). I don't think it has gotten any better since then, do you?

High levels of methane gas and CO2 are generated by the rotting rubbish in the ground. These are greenhouse gases, which contribute greatly to the process of global warming. Many toxic substances end up in a landfill, which leech into the earth and groundwater over time. This creates a huge environmental hazard. (forgeskiphire.co.uk/blog/landfill/)

Globally, we're producing a colossal **1.3 billion tons of landfill waste** annually, with a projected increase to **2.2 billion tons** by 2025. Despite efforts by many of us to "Reduce, Reuse and Recycle", the fact is we're generating more landfill trash now than ever before. Worse yet, our waste is projected to nearly <u>double</u> globally over the next 15 years.

Today, the average American throws out about 1,000 pounds of garbage each year. Americans generated about 250 million tons of trash last year, according to U.S. EPA estimates. (steelysdrinkware.com/growing-global-landfill-crisis/)

The worst part is the fact that we are also sending a lot of garbage to other countries. A Guardian investigation has found that hundreds of thousands of tons of US plastic are being shipped every year to poorly regulated developing countries around the globe for the dirty, labor-intensive process of recycling. The consequences for public health and the environment are grim. (theguardian.com)

Did you get that part? We are sending our waste to other countries because we are trying to circumvent our own regulations. And how do we send it to other countries, you might ask? Well, would you be surprised if I told you that we stuff that crap into containers, load it onto container ships and then we send those bunker fuel toxin gushing boats off to some far away place, so we can tell ourselves that we solved the issue, we threw it *away*.

Can you guess what country participates in accepting western plastic waste shipments? Yes, of course, Bangladesh is one of them. They are so poor, because the rest of the world doesn't give a hoot, that we started to exploit them even more, by poisoning them with our garbage. They are drowning in our plastic garbage in "away" country. We are just glad in the United States that Bangladesh is so poor and is so far away, when their country sinks into the ocean, they can't afford to fly to the US, so we won't have to deal with their global warming refugees... oh yeah, we dodged that bullet.

The U.S. Census Bureau recently published complete 2018 export data for shipments of plastic waste (officially called "waste, paring and scrap") generated in the U.S. and sent to other countries. 78% (0.83 million metric tons) of the 2018 U.S. plastic waste exports were sent to countries with waste "mismanagement rates" greater than 5%. That means about **157,000 large 20-ft** shipping containers (429 per day) of U.S. plastic waste were sent in 2018 to countries that are now known to be overwhelmed with plastic waste and major sources of plastic pollution to the ocean. The actual amount of U.S. plastic waste that ends up in countries with poor waste management may be even higher than 78% since countries like Canada and South Korea may **reexport** U.S. plastic waste. (plasticpollutioncoalition.org)

How humiliating is it for the USA to have to admit that we are incapable to take care of our own trash? We are the dirtiest pigs on this planet.

If you like to learn more about the global waste trade, a good place to start is here:

https://en.wikipedia.org/wiki/Global_waste_trade

It gets better. Indonesia has also been found to "re-export' illegal waste to other countries....so we send it to Indonesia, thinking our problem was solved, but then they send it off to yet an even poorer country, and thus we all contribute to the demise of countries like Bangladesh, and none of us want to help when 30million refugees will try to escape our garbage and toxins, that created global warming, that led to sea level rise and the drowning of a country. Luckily, Bangladesh, like so many other affected countries, they are all far *"away"*.

There is a very disturbing picture on the internet, of women working in the landfill, sourrounded by toxic and stinking trash. The women have woven baskets on their backs, a stick in their hands and cloth over their noses and mouths. They work in the landfill. Yes, they WORK there. They are digging in our trash, they are digging for something......

......trying desperately to find anything that could be valuable, maybe something could be worth a few pennies, and if you work 10hour days in the stinky toxic landfill, you could earn a few bucks. Why don't we send them some more of our garbage? I have a better idea, let's just send our government officials there so they can help these women find treasure, and maybe we can add some CEOs from our oil, coal and gas companies as well.

If we just had enough trees on our planet, maybe they could absorb some of that CO_2 and help us prevent more damage from global warming, that could be an idea, right? Here goes the naïve human race, the way of the dodo bird. I don't think you have seen what we do with trees on our planet. Let me tell you:

From **2001** to **2019**, there was a total of **386Mha** (million hectares) of tree cover loss **globally**, equivalent to a **9.7%** decrease in tree cover since **2000**. (globalforestwatch.org)

We have an ever-growing population, an ever-growing pile of garbage, ever-growing numbers of pollution and toxins, and we have less and less resources every year. When climate models look at industrial CO_2 output, and they count for all boats and cars and planes etc. how many of them also take into account the fact that we have less and less trees to absorb the ever-increasing numbers of CO_2? Hardly any.

More and more people produce more and more CO_2 and garbage and need more and more food, that creates a need for more agriculture. Those trees are in the way. Let's cut them down, forests be damned.

Between 1990 and 2016, the world lost **502,000 square miles** (1.3 million square kilometers) of forest, according to the World Bank—an area larger than South Africa. Since humans started cutting down forests, 46 percent of trees have been felled, according to a 2015 study in the journal Nature. (nationalgeographic.com)

Worldwide, 19 percent of cities with more than one million inhabitants are already located in areas with a high to very high risk of drought - a total of around 370 million people are affected. Aug 25, 2019. (panda.org)

Here is a tipping point from National Geographic.
The levels of atmospheric CO_2 are rising, and it's assumed that eventually, plants won't be able to keep up.
"The response of the land carbon sink to increasing atmospheric CO_2 remains the largest uncertainty in global carbon cycle modeling to date, and this is a huge contributor to uncertainty in climate change projections," the Oak Ridge National Laboratory notes.
Clearing land for ranching or agriculture and fossil fuel emissions are the biggest influences on the carbon cycle. Without dialing those two things way back, scientists say a tipping point is inevitable.
"More of the CO_2 we emit will stay in the air, CO_2 concentrations will rise quickly, and climate change *will occur more rapidly*," says Danielle Way, from Western University.

This from ran.org, ...as for how many trees are cut down each year, *IntactForests.org* concludes that intact forest landscapes from

2000-2013 were reduced globally by 70,000 square kilometers per year (about the size of Costa Rica) for a total of 919,000 square kilometers. As to the "number of trees" this represents, it's impossible to get an accurate count. Tree density in primary forests varies from 50,000-100,000 trees per square km, so the math would put this number at **3.5 billion to 7 billion trees cut down each year.**

Rainforests across the world are in great danger. Food and Agriculture Organization's 2016 State of the Forests report revealed that 7 million hectares of forest are lost annually while agricultural land expands by 6 million. The biggest threat to forests today is industrial agriculture production of commodities like Conflict Palm Oil, fabric, paper and logging. Only 4 billion hectares of forest remain worldwide according to Global Forest Resources Assessment 2015.

Do you really think we even have 50years left, if that many at all, before we see a total collapse of our own environmental systems and habitats? Not if we keep cutting like this.

Now let's build some sandcastles. Please watch the movie Sand Wars.

Yep, you need to know about sand. You see, concrete requires sand. Building anything with concrete, roads, houses, skyscrapers, etc. it all requires sand. Problem is, we don't have enough, at least not the right kind. There is a big difference from desert sand to beach sand.

Desert sand has little use; the grains are too smooth and fine to bind together, so it is not suitable for the making of concrete.

We literally have tons of it on beaches, deserts, and under the ocean, but we're using it up faster than the planet can make it. We use sand way more than you'd expect. Worldwide, we go through **50 billion tons of sand every year**. That is twice the amount produced by every river in the world. <u>After air and water, sand is our most used natural resource</u>. We use it even more than oil. It's used to make food, wine, toothpaste, glass, computer chips, breast implants, cosmetics, paper, paint, plastics. *(businessinsider.com)*

And how is this sand transported, may I ask? Can you guess? See any big boats around here?

The world is running out of sand. Eearth.org.

Let's try to see the connections here. We are cutting down more trees than ever, spewing out more pollution than ever, generate more garbage than ever, we require more food than ever, we have less clean fresh water than ever, and our planet is getting warmer at an alarming pace. All that leads to global warming refugees, who will have to move, migrate, escape, whatever you want to call it. Then we will need to build new cities for the millions and millions of people that are running from starvation and collapsed countries, with just one problem: We won't have enough sand to build those cities!?

When the last tree has been cut down, the last fish has been caught, the last river has been poisoned, only then will you find out that you cannot eat money.

Fugitive emissions from energy production cause 5.8% of our global emissions.
...fugitive emissions are the often-**accidental leakage** of methane to the atmosphere during oil and gas extraction and transportation, from damaged or poorly maintained pipes. This also includes flaring – the **intentional burning** of gas at oil facilities. Oil wells can release gases, including methane, during extraction – producers often don't have an existing network of pipelines to transport it, or it wouldn't make economic sense to provide the infrastructure needed to effectively capture and transport it. But under environmental regulations they need to deal with it somehow: intentionally burning it is often a cheap way to do so.
Fugitive emissions from coal (1.9%): fugitive emissions are the accidental leakage of methane during coal mining. (ourworldindata.org/ghg-emissions-by-sector)
"*Fugitive*", the criminal. I don't think that this gas is the criminal, I believe the companies that built this stuff are the criminals, what do you think? And what do our politicians think?

That is just the conveniently and accidentally released amount of toxins we spew out in the United States, just by 'accident', because it is easier and cheaper this way.

How much do you know about fish?

CHAPTER 8

This year alone, 2020, for the first time, Turkey became the largest refugee-hosting country worldwide, with 1.59 million refugees. Turkey was followed by Pakistan (1.51 million), Lebanon (1.15 million), the Islamic Republic of Iran (982,000), Ethiopia (659,500), and Jordan (654,100). Those are refugees from wars that were intensified and mostly triggered because of global warming.

According to the UN, an estimated 362,000 refugees and migrants risked their lives crossing the Mediterranean Sea in 2016, with 181,400 people arriving in Italy and 173,450 in Greece. In the first half of 2017, over 105,000 refugees and migrants entered Europe.

How many refugees are there around the world? At least **79.5 million people** around the world have been forced to flee their homes. (unhcr.org)

If you think about connections, it is no wonder that many countries around our planet want to destroy America and everything it stands for. How many of your fellow brothers and sisters do you want to watch living off of our garbage?

Not only do we destroy more of our natural habitat on Earth than any other country, but we don't even try to help other countries to mitigate the damages that we cause, never mind that we never pay them for it, but we destroy their economies in the process, that creates wars and global warming refugees.

A lot of those people are fleeing droughts and the lack of food and fresh water. Sadly, many coastal African countries are unable to fight big international conglomerates that send giant fishing and processing boats to their coasts.

Overfishing is a huge problem. It can change the size of fish remaining, as well as how they reproduce and the speed at which they mature. When too many fish are taken out of the ocean it creates an imbalance that can erode the food web and lead to a loss of other important marine life, including vulnerable species like sea turtles and corals. (worldwildlife.org)

Overfishing can wreak havoc and destroy the environment and marine ecology and completely disrupt the food chain. For example, herring is a vital prey species for the cod. Therefore, when herring are overfished the cod population suffers as well. And this has a chain reaction on other species too.

Apr 9, 2014. (marinesciencetoday.com)

Never mind that big fishing vessels take ungodly amounts of fish out of the ocean with just one run of their gargantuan nets, on top of all of that we destroy the areas where the fish live, or, well, used to live. We destroy their habitats with acid rain from our factories that falls into the oceans and increase the erosion of plankton in surface waters. Smaller fish live on plankton, bigger fish live on those smaller fish, Dolphins live on those, etc.

Habitat loss, releasing of toxic materials, and overfishing are the three greatest threats to aquatic biodiversity. No species can survive without a habitat. (quizlet.com)

As if it was not enough that we kill our planet, but we also speed up the process by fishing out whatever is left, as fast as we can. I wonder what the last Tuna fish will bring at the last auction of actual real fish? Maybe a few million dollars? I bet it will be worth it.

It has been estimated that between 1 to 2.7 trillion fish are caught from the wild and killed globally every year: This doesn't include the billions of fish that are farmed. Jul 24, 2017. Forbes.com

A team of scientists led by David Kroodsma from the Global Fishing Watch published a paper that put the figure at 55 percent—an area four times larger than that covered by land-based agriculture. Theatlantic.com, September 2018

Unless humans act now, **seafood may disappear by 2048**, concludes the lead author of a new study that paints a grim picture for ocean and human health. ... The research also found that biodiversity loss is tightly linked to declining water quality, harmful algal blooms, ocean dead zones, fish kills, and coastal flooding. Nov 2, 2006. NationalGeoGraphic.com

You see, not only do we destroy their habitats, on top of that we catch as much fish as we can possibly get our hands on, like there is no tomorrow......and then we add the fact that over 1 million marine animals (including mammals, fish, sharks, turtles, and birds) are killed each year due to plastic debris in the ocean.

Let's imagine where this is going: you are young boy, 11 or 12 years old, you watch your parents struggle every day to bring food to the table. Your father goes out on a small wooden boat, every day, with the other dads, trying to catch some fish. You haven't been to school in 2 years, you had no time to go. You had to help your mother every day to walk further and further into the hot land, trying to find firewood to cook, in case your father might come back with a fish, or two.

But the days where the family could count on catching fish were long gone. What used to be larger fish, a boat full of them, and happy smiling proud fathers bringing their boat back on shore, has turned into bitter disappointing and painful years. Many families had left, your friends had gone, many fathers had left their families, trying to make their way into Europe. If only they could get to Europe, they could live. Even if they could not find a job there, at least those countries would not let them starve to death.

If only, if only you were old enough to work on one of those big boats, maybe they would let you have a few fish, so you could take them home to your mother and stop her from crying and starving.

Then the day comes, your father has decided, the family will leave this rotten place and join a few others, they will leave and go to Europe. You had heard of it, but you didn't really know what another country was, or even what Europe was or meant. But if your dad said we shall go, then it was decided. You packed whatever small belongings you had and held your mom's hand for a while when you walked away from your village.

The trip took many weeks, you ate berries that you picked with your mom, sometimes dad brought some food. You never asked where he went, or how he got the food, you just knew it had something to do with the blood stains on his shirt.

After a long trip, full of pain and horrors, your mom smiled again, hope was near. There was another ocean in front of you and many people at the beach, all trying to find a place in a larger boat. You had never seen so many people in one place, all trying to get on the same boat. How your parents were able to get a spot on the boat was unclear, you were just happy to see your mom smile again, after all the things that the tracker men did to her on this long journey.

The boat was sturdy, packed with people, it smelled, many had thrown up from sea sickness, and you were just out of the bay, you had not even started to hit the big waves yet. You held on for dear life, literally, as you had seen already 2 people that fell out of the boat. Nobody tried to rescue them, and everyone pretended not to hear their screams.

That was the past. It is now 8 years later, you had arrived in Spain, your family made it, and you ended up in a camp with thousands of other refugees. You were an outsider, you were shunned, disrespected, spit on, beaten and treated like an animal in any city that you went to. Finally, you had found some work, hard construction work, were you had to shovel some kind of wet dark dirt into steel drums. It smelled really bad and you had a headache every time you worked there, but at least they gave you some cash money at the end of each day. It was not much, but for a guy who was still learning to speak the language, it was all you could get.

One day, when you returned to your refugee camp, your parents were gone. Their clothes and bags were still in the tent, but they were nowhere to be found. You waited a few days, but they never came back. Given the awful tasting pasty food that was being given to refugees, you had no reason to stick around. Besides, that cute girl from the shop you passed by ever so often, she kept smiling at you and you could actually understand each other.

Now, now you are a tourist from America, enjoying Paris in the evening, strolling around with your family on a long vacation, taking in the scenery, maybe, just maybe, the screeching tires from the van make such a loud noise that it stops you dead in you tracks. You grab your wife and child and try to rush them into the nearest café entrance, where all the other patrons are trying to rush in and escape to safety. But too late, as you stumble and fall you can hear the rapid-fire sound from the AK-47 as the passenger of the van opens fire at the café. Yes, our little boy grew up and he had learned how to fire an assault rifle. He was trained to do this. All he had to do was learn how to shoot this rifle, and in return they would give him food, water, clothes, shelter and be his friends. He was so glad he finally made it to Europe. He never did find his parents again.

Ironically, you thought about ordering some nice tuna steak for dinner.

All of that over fish. The toxin spilling boats, the inhumane conditions of working on those boats and the total destruction of marine life on our ocean floors... all of it, just for fish and money.

When there is a swarm of locust descending down on a field and eats it all up in a matter of hours – does that swarm of locust know that they are locust?

CHAPTER 9

Have you considered the additional extra 100 million people we add to our planet every year? That is 100 million more, never mind the few that die, this is the EXTRA amount, after we count out the ones that died. 100 million people more, every year. Guess what, they will need raw materials, and those raw materials need to be transported to wherever these new 100 million people will live, preferably in that shiny new city you built for them. How about we use bunker fuel container ships to transport all of that? Yeah, let's do that. That's a great idea, our new container ships sail much slower, so they won't pollute as much while they transport these extra 1billion tons of materials. (that's why we need more boats than ever before)
And how about all the garbage that these extra 100 million people will produce, year after year after year, what about the fresh water they will need? 100 million people that will need products, they need to be manufactured, that causes even more CO2 to be released. OMG, are any of those 100 million people going on a cruise? Let's send out an extra 100million Carnival Cruise brochures.... yeah, let's cut down some trees to print those brochures.
Does the newest and latest prediction model consider that we will lose an extra 2billion trees this year, and how that will affect our CO2 absorption, by the time these extra 100 million people will start breathing?

When a prediction model says there will be a gradual increase in temperatures and sea level rise because that is how it all started, they are wrong. Each year it increases a bit more than it did the year before, each new year there is just a bit of an increase over last year's increase. Now project that out to the next 30 – 50 years, and voila. You end up where most models say we might be in 100 years from now. That is the problem. Everybody says: "Oh, hundred years from now? Damn, I won't be around, why should I care?"

We all have what is referred to as Confirmation Bias. Oh, how I wish I could confirm my hopes, how I wish I could find something that I agree with, that will confirm my inability to contemplate that it will be a lot worse than we all thought, and that this will happen a lot sooner than we all hoped. We can't think about the end, because in our mind, the end is a terrible idea, so let me find more data that will confirm that we are not at the end, yet, and then maybe I can be a happy scientist and I do not have to explain this stuff to anyone, or, god forbid, I have to look my family in the eyes when I pull the trigger?

Can you imagine, in the year 2035, during an extremely hot summer in Washington, D.C., the president addresses the nation, and the planet, in an impromptu breaking news conference:

"My fellow Americans let me get straight to the point. Our scientists, along with over 2000 scientists from around the globe, have confirmed that we only have about 50years left on our planet, before we run out of fresh water, crop land and breathable air. We need to start immediately to shut down all non-essential sources of pollutants, factories, cars, boats, planes and businesses that are not needed for the survival of mankind......"

Where is the model that shows the additional 100 million people this year, another 100 million people next year, and the next, and the next, and the next after that? Even if google says we only add a little over 83million people per year, that still means that in 5 years from now, over 400 million additional people would like to breathe, bathe, and eat some fish, if possible, while burning some wood in the oven, and after that dinner, they will throw their garbage away, without having to pay some huge fee for their rubbish.

Where are they supposed to live, what cities are being build for them? Are they all going to work in coal mines, on an oil rig or on fishing boats? They will probably consume about 4 Billion Tons of raw material per year. How are we going to transport all that? All of that is in addition to the damage we already caused on our planet.

We have less resources every year, less clean air, less fish, more pollution, more toxins, more garbage every year, and we are adding more people every year that cause us to search for and dig up more resources every year and manufacture even more products than the years before.... this growth is killing us.

When a logging company says that their business is booming, that is a HORRIBLE thing to say. It means they are cutting down more trees than the previous year, transporting more crap, more CO2 than the year before, more resources destroyed, more pollution, more, more, more, more .... and none of that 'more' has any positive impact on our lives or planet. It just brings us that much closer to the death of our children.

We have too many people on our planet already.

Babies are cute, yes, but they also need to eat, they need clothes, they need water, later on they want to have an Xbox and all that stuff needs to be produced and shipped and all that is killing us... yes, babies are killing mankind. Who has a plan for that?

"It is baffling that we willingly and knowingly continue to sow the seeds of our own destruction," said UNDRR chief Mami Mizutori and Debarati Guha-Sapir of Belgium's Center for Research on the Epidemiology of Disasters, in a joint foreword to the report.

"It really is all about governance if we want to deliver this planet from the scourge of poverty, further loss of species and biodiversity, the explosion of urban risk and the worst consequences of global warming."

But researchers warn that "the odds continue to be stacked against" these communities.

"In particular by **industrial nations that are failing miserably** on reducing greenhouse gas emissions to levels commensurate with the desired goal of keeping global warming at 1.5 C as set out in the Paris Agreement," Mizutori and Guha-Sapir said.

They called on countries to do more to strengthen disaster risk governance and to better prepare for future climate catastrophes. **Currently, the world is on course for a temperature increase of 3.2 degrees Celsius or more**, unless industrialized nations can drastically cut their greenhouse gas emissions.

That projected temperature increase is enough to increase the frequency of extreme climate events across the world, the report said, rendering any improvements to disaster response or climate adaptation "obsolete in many countries."

"We have seen little progress on reducing climate disruption and environmental degradation," said UN Secretary-General António Guterres. "To eradicate poverty and reduce the impacts of climate change, we must place the public good above all other considerations."

(undrr.org)

Did you catch the part where we are well on our way to increase our planet's temperature not by just 1.5 degrees, but by 3.2 or more, unless we drastically stop or reduce pollution?

What's the difference there? Well, it is either New Orleans has a few more floods every couple of years, or New Orleans is 2-5 feet under water, permanently. Same for New York, Miami, San Francisco, London, Bangladesh and just about every coastal city on earth. No big deal, right?

We cannot talk about industrial pollution and toxins without talking about our destructive population growth. You can't fix one thing while ignoring the other. Please, wake up.

Now think about the fact that we will have to move millions of people from Florida and other low-lying areas (Louisiana) and the likes to higher ground. Will they all try to move to Iowa? Is that State ready? You got the roads, the water pipes, the food supplies, the stores and the housing, all that is ready to go? No? Well, what are you waiting for? If you don't start building now, you will never catch up to the coming problem.

Do we even have enough land to grow the food that all these people need? No, we actually do not. Yeah, read that again. We do not have enough room to grow enough food for all the humans we currently have on our planet. What are we going to do when we will have an additional 1billion people, all looking for food, fresh water and fresh air? Where are they supposed to put their garbage? Do you think our landfills are big enough to handle all those extra billion tons of rubbish?

Let's look at the land and how we use it, in the United States. Many people have never seen this map, you can find it easily on Google. Search for land use map USA. You will see an astonishing big giant yellow blob, right in the middle of the map of the United States. That giant blob, that is the area we use for raising cows. Yes, cows. That area is bigger than Texas, New Mexico, Arizona, then goes up north to cover Utah, Colorado, Kansas and crushes half of Wyoming, Nebraska, and Iowa. Yeah, that is just the area we use for cows. Then we use an area twice the size of Florida, just to grow the food for all those cows.

All that creates methane farts for our atmosphere and incredibly stinky and dirty run-offs into our rivers, an endless zest pool that just gobbles up every ounce of water they can find. We grow all that area of food, to feed it to the cows, to then kill the cows and eat them. That really doesn't make any sense anymore.

Why do we even have to create laws to prevent people and companies from polluting our environment? Don't they understand that you cannot poison our environment and then expect to live in it?

What does the Pentagon have to say about all of this?

CHAPTER 10

Here is something to think about. We knew all along, but we opted to spend money on killing people, killing each other, rather than on saving mankind. And we have gotten really good at that killing thing.

'Climate change is a national security threat. Stronger storms will lead to increasing damage to coastal military facilities, as when Hurricane Michael caused substantial damage at Tyndall Air Force Base in Florida. Stresses on resources and agricultural changes will increase the global flow of refugees and cause cross-border instability. That, in turn, will mean greater involvement of U.S. forces around the world.'

For many years, the Pentagon included the impacts of climate change in our defense planning. In **2003**, the Department of Defense under President George W. Bush concluded climate change "*should be elevated beyond a scientific debate to a U.S. national security concern.*" The last several Quadrennial Defense Reviews, which assess the array of threats facing the United States, have considered the impact of global warming. The QDR stated that "*the impacts of climate change may increase the frequency, scale, and complexity of future missions.*'" (militarytimes.com). GREAT! In other words, we need more money for more bombs, big battle cruisers, missiles, and other ways to kill other humans that are just trying to survive the drought and heat and lack of water etc..... let's kill 'em all!!! Is that the plan?

You really can't come up with a better plan? If that's the best we can come up with, then we're totally screwed.

So do we spend money on building cities for all the new people and the ones that will become refugees, or do we rather spend our money on more and more military killing machines to fight the people that try to run away from famine disasters? I mean, come on, it's not going to be a big fight, is it? Most refugees do not try to escape from pollution and global warming in a fighter jet, are they? How badly armed can they be? Are you really going to shoot and kill all the refugees from Louisiana and Florida that will try to flee to Iowa?

And why are we focusing our money and military on the outcome and result of global warming, instead of working on *preventing* this potential outcome? You need a military to do what exactly? Defend the polluted United States, the toxic rivers, the dry desert that used to be Texas, the forests that are no longer there? NOBODY wants to attack that crap and take it away, we are killing it ourselves, we are poisoning it, all of it.... who do you think wants to come over here and get it? Nobody. Then why do we need ever more military forces to defend an ever-dwindling stock of supplies and resources? What are these '*enemies*' going to do, take away our landfills? How is an aircraft carrier stopping methane gases from floating up into our atmosphere? How is a battleship going to re-freeze the tundra? That shiny new tank is going to remove the acid from our oceans? Really?

The US military is protecting Oil fields in Iraq and poppy fields in Afghanistan right now. Can you not see armored personnel carriers and even tanks protecting freshwater aquifers in the United States in 20 years from now? Can you not see US Soldiers standing guard over the last remaining forests? Is that the future you really want? Really?

Now let's get to the root of the problem, the actual criminal polluters. Cleaning up the toxic mess that your company creates everywhere will require money to pay for the cleanup. We will charge YOU for creating the mess and that will pay for our cleaning costs.

When we finally tax oil, gas and coal to account for the environmental damage that they cause, plus the healthcare costs for all people affected by these unfiltered poisons, maybe then we can have an open and honest discussion about the "economics" of renewables. All alternative energy sources are really much cheaper and much more economically viable than anything else on earth!

When more than 90% of scientists tell you that it is a very bad idea to dump your toxins into our rivers, that is a fact. We do not need the other 10% to agree anymore.

"But I'm telling you man, the scientists don't know everything dude...."

"How much?"

"What you mean?"

"How much, how much time did you spend studying this?"

"Huh?"

"How many hundreds of hours did you spend studying this subject matter before you formed your highly educated and experienced opinion on the statement that toxins in our environment are not a bad thing?"

"...well, uh, but not all of them agree.... "

"It doesn't matter, enough of them agree, and you really don't need to be Einstein to comprehend that more toxins can not ever be a good thing, right?"

"I guess........."

"And if you never studied any of this, if you didn't spend at least 500hours on it, then you really don't have enough base knowledge to form an opinion, do you?"

"Well......"

"Then why da heck do I need to create all these laws to make a decent human being out of you so you will stop killing our children?! You dumb morons always go on and on about less government, and less interference, and our freedom to do whatever... well, if you weren't so damn stupid and you'd behave without government interference, we would leave you alone, gladly so. But you are just too freaking stupid to comprehend basic issues... like not dumping your toxic crap in our rivers......" Damnit, people are brainless.

If I were to dump a truckload of oil infested sludge onto your front lawn, would you just shrug your shoulders and say: "Eh, what you gonna do?" and just go about your day? Of course not. But if I dump it in the nearby river, you have nothing to say about it then? Why is that?

Then why are we not doing anything about polluters in our own country?! And isn't the fact of inaction a borderline negligent homicide? "Oh, yeah, I know I was polluting, lots and lots, but you know, there was no consequences, not really, I got a letter from some city clerk, but that's about it. So what if there is a school nearby, not my fault. Some kids might die because of the sludge I drained into those pipes? Really? Nah, you can't prove that; I'll just keep doing it...nobody is trying to stop me........"

That is the attitude that many people have. How is it possible that we let them get away with that? It's not about more government and more police – but if that's what it takes, then so be it. Obviously, many toxin terrorists are not going to stop on their own just because we asked them to be nice to our environment. We cannot allow others to destroy our lives just because they want to sell us cheap products, but they refuse to pay for the damage that their factories create, so they pass those costs on to us when we have to take our child to the emergency room. Why do you want to keep paying for their damages?

We cannot allow a few slimy operators to reap all the benefits and then pretend they didn't have to clean up their mess.

This inaction on behalf of our elected officials is, in and by itself, proof enough that our politicians are all corrupt. A bold statement? I don't think so. What possible reason could you provide why we should continue to dump toxins in our water, air, ground etc.? Please, provide one reason. Ok, please, provide one GOOD reason.

We cannot say that we need to protect the jobs when those very jobs are the reason why we are in so much trouble on our planet. What good is it when your job literally contributes to the death of your own children?

Are we going to talk about more taxes? Oh yes, yes, we are and you gonna love it. Read on.

CHAPTER 11

Taxing air pollution, or any pollution, will certainly change the accounting dynamics that we allow to destroy our home planet. It is that accounting method that forces some companies to pollute, because they cannot afford not to pollute. Their profit margins are too small, ergo, their product is cheap, and we end up paying for their pollution in more ways than one.

Some will say that this idea is equal to a draconian dictator government, stay out of my business etc., you can't tell me how much CO_2 I can put in the air, there is no proof that I hurt anyone....blah, blah, blah.

If you could operate a clean business, without polluting our environment, then there would be no need to tax the pollution. But if you can't figure out how to run a clean business, then I have to elect new leaders that will protect my life and the future of my children.

How about we make every CEO drink the crap that comes out of their factory pipes, mix it with Cool Aid and have their kids drink it, then pump their house full of the junk that comes out of their factory smokestacks? Yeah, eat, drink and breathe your own company poison, for a few days or weeks, then tell me how clean your company is. Let's put that on a reality TV show, which ever company contestant survives the show after 10 episodes of toxic death wins the beautiful new set of steak knives.... (this is a movie idea, contact me to buy the rights)

How long do you want to wait until we impose real strict laws against pollution and toxin terrorists in our own country? Shall we wait another ten years? Why? How bad does it all have to get before you have the balls to create meaningful and tough penalties for polluters, and how long do you want to wait before you start taxing the real costs of toxins? How about we wait another 15 years, will that be long enough for you? In 2035, when we have another 1.2billion people on our planet, when our country experiences the 6th major drought on record? When you can only swim in the last 25% of our lakes? Just give me the time frame. I just want to know WHEN do you think we need to make some drastic changes. How much damage, destruction, carnage and dead kid's bodies riddled with cancer do you need to look at before you say it cannot go on like this?

Why are we prolonging the inevitable stricter laws that we all know we will have to implement, sooner or later. Why are we waiting so long? Why are we waiting at all? Do you truly, honestly believe, even for a second, that it will get better? On its own? By itself? Really, suddenly all these polluters wake up one day and say: "Darn, maybe I should not pollute anymore, I made enough money, I can start cleaning up after myself...." When do you think that will happen? Ask yourself, honestly, do you think that will *ever* happen?

Let's talk about BUGS! Yes, critters, crawly little nasty things, locust, etc. You are already eating them, probably every day, you just didn't know it, yet.

(*sustainableboost*.com)

No way you'll eat bugs, you say? Well, sorry to break this to you, but if you eat chocolate, pizza and spaghetti, you are ALREADY eating insects -- and worse.

The US Food and Drug Administration allows 30 or more insect parts and some rodent hair in every bar of chocolate; nearly two maggots in a 16 ounce can of tomatoes or pizza sauce; and up to 450 insect parts and nine rodent hairs in every 16 ounce box of spaghetti.

There's just no way to get rid of all the creatures that might hitch a ride along the food processing chain, so the FDA has to allow what they call "food defects," which you eat without knowing. They are in many of our foods -- even peanut butter and jelly.

Your friends and neighbors are already munching on insects. According to a report by Global Market Insights, the US edible insect market topped $55 million in 2017 and is expected to grow to nearly $80 million by 2024.

Europe's on track to do the same, while Asia Pacific nations are expected to eat $270 million worth of insects by the quarter of the century. A growing demand for high quality protein, along with a movement toward sustainability and against processed foods, are a few of the reasons behind the growing popularity.

Here's another explanation: Bugs are very good for you.

Many countries and traditions have known this for decades, even centuries. According to a 2013 report by the United Nations Food and Agriculture Organization (FAO), at least 2 billion people worldwide eat bugs every day.

How is that saving our planet? Easy. It cuts down on fish and meat/poultry for protein, it saves farmland that is being used just to feed cows and pigs. It cuts down on the area used for farming animals, it saves on methane gas and run-off pollution. And NO, you won't know the difference from bug flour to wheat flour. If

Russia had a large bug to flour operation, they may not have had to pull their wheat from the global export market, which could have saved the Arab Spring.

If you look at the resource costs for a pound of beef, you could achieve a conversion rate of 20 to 1. That is one pound of beef comes out of 20pounds of feed. You have to feed the cow 20 pounds of food, for every one pound of beef that you could get out of the cow. For bugs, that ratio is just 1 pound of protein for 2 pounds of feed. And the feed that these bugs eat is a lot easier to farm than feedstock for cows. It can take months to feed an animal to get it ready for market and slaughter, while you could be on your third harvest for insects already.

No worries, these little critters are flash frozen, you will never see a cricket leg in your tortilla or sticking out of your toast. However, for the devastating effects of global toxins, overfishing, water shortages and an ever-increasing amount of pollution and droughts, well, this is one of the best ways to work on our food security.

If Egypt had enough of these bug farms, they would have had food security, or at least flour security, they would not depend on Russian imports for wheat, they would not have riots etc...

When we start to tax the pollution, Cruise Lines will think very hard about their business practices, and some simple sensors on board could let us all know how much crap they put out. Then they get taxed accordingly.

If the sensors onboard your cruise ship show an elevated amount of emissions from your boat, you are not allowed to dock in our harbor. If the sensors show that you dumped any kind of garbage or sewage in our oceans, your operating license will be suspended, immediately, and your company will be fined $300,000 per day that this boat was operating outside the sensor limits. Don't pay that fine every day and your toxic boat will be seized, your stock will be delisted, and or trading will be frozen. We should add a toxin tax to all cruise ship tickets, right from the start.

Why do we allow companies to get away with minor fractions of a fine?

There is no good guy, model citizen, a pillar of their community, none of that is true if the CEO is a polluter.

Polluters are criminals. They are not businessmen, they are not the good guys, no matter how big their company is and no matter how rich they claim to be. They can't be honest, that's how you know they are criminals.

"Hi, I'm John, I own a little bakery in town, what do you do?"

"Oh, hi, I'm David, I am a cruise line operator. We charge people around $600 to participate in our toxin dumping journey. On average we kill about 20-30 dolphins per trip, destroy about 30-40 square miles of ocean floor habitats and all the crustaceans that used to live there. Of course we stop pumping out all that raw sewage once we get close to a harbor, you know, you really don't want to smell all that shit near you before you go on board, ha-ha, you know what I mean?!!" he raised his hand for a high-five.

Really? Do they ever tell the truth?

Carbon credits do have merit, if they are applied properly and more controlled, they can work. At the least they force companies to help in reducing the pollution that they put in our environment. Not all of it, not most of it, but some of it.

But all these solutions have to come with real consequences. No carbon credit, no plane ticket. No sensors on the smokestacks, no cruise ship vacation. It cannot be that we allow companies to pollute, and let's face it, WE ALLOW THEM to kill us. We are literally sitting at the bank of river, watching the other side, and watching how they roll barrels full of toxins and contaminated crap, one after the other, in the river. We just sit there and watch them do it. Maybe, just maybe, somebody sits there and takes notes, or even pictures. But so what? The barrels are still rolling off that truck right into our river. Maybe, maybe in 5 years from now we can be in court and talk about it, but that doesn't stop what already happened, and it doesn't stop what happened in the 5 years since then.

However, if we tax them based on their pollution, and they have to pay for it BEFORE whatever product they make gets to market, it will make them pay attention. They will need to figure out how to pollute less. They have to figure out how to make products with a lot less pollution, and they already know how to do that. They just don't want to, they don't have to, because you keep voting for corrupt money hungry politicians that won't enact any laws that will stop these criminal enterprises and toxin terrorists from killing our children. Only if they pollute a lot less or not at all should they be able to avoid the costs of cleaning up their mess.

Now, let's talk about wind, not all wind generators are those large, clumsy, heavy, giant poles with a generator on the top and big propellers flopping in the wind. Not the most efficient to erect and build. Big problem, they can only take wind from one direction. Big advantage, they are tall, where the wind is stronger, they do make some noise, and yes, some birds fly into them. But they don't kill any more birds than cats or windows. However, they are not the most efficient design. Great for offshore wind farms, or in large valleys with lots of wind that usually comes from only one direction. In certain places, these super large ones are the best, yet.

There are other, and may I say, better designs, that are more efficient, don't kill birds, easier to erect, easier to build and also easier on the eyes. Yes, they do exist.

The completely silent and more efficient designs are vertical or cone-shaped wind generators, that do not have any windmill propellers, at all. Their generator is on the bottom, easier to build, easier to erect, easier to maintain. And your city or government officials have probably never shown you any of these, did they? Please search for vertical wind generators.

You can place these really close to each other, birds can easily see them. You can place them almost anywhere and with a little breeze, you have clean energy. Why is it that California has almost a thousand miles of state-owned beaches, and none of those have any wind generators on them?

There are *bladeless* wind generators as well, yes, NO blades, no visible moving parts, none at all, they look like a thick stick in the field, they also don't make any noise. Which one of these should we use? Well, you may have guessed it, we need to use all of them. Yes, every design has merits, we need to fund ALL those projects. Some of them side by side. How much wide-open windy prairie land do we have that we are either not using, or won't be using anymore in 5-10 years anyway, because it will be too hot to grow anything on those lands? There is plenty of empty space throughout our country. We have torched our planet already, let's at least get some use out of the burned-out areas.

"But I don't want to look at those ugly wind generators in my back yard...."

Really? A wind farm doesn't sooth your eyes? Your kids dying of heat stroke and lack of food looks better?

Oh, no worries, you won't live here much longer. You won't have to look at them for long. Or, maybe, we can just dump all that coal ash in your backyard since you seem to prefer the destruction of our land over saving our lives.

Look, we are not putting them in your back yard. However, if you are planning a new development, you should be required to install enough sustainable energy projects to support the energy needs of your finished project.

With that, every new house or building should be required to supply at least 50% of their own energy needs, or you are not allowed to build it. It's simple. You want to build a house, that house needs to have enough solar panels in its design to provide at least 50% of the energy needed to run that house, without ever plugging it into the city grid. If your house/design cannot do that, then you need to get back to the drawing board. Maybe you are not that good a designer, maybe you need to learn more about alternative energy, whatever it is, you are *not* ready to build a house if you cannot take care of your own energy needs.

No wind or solar in your house plans? No permits. No construction. If you don't care to help provide the clean energy needed to run your house, why should we care to provide you with the permit or license to build another environmental disaster in our town, city, county or State? Go away and try to live in another country.

Hey, I heard that beachfront property is really cheap in Bangladesh, maybe you can move there?

Now let's talk about money.

CHAPTER 12

You have to learn a bit about our money and where it comes from, because politicians will keep telling you that we just don't have enough money to save our own planet from ourselves.

Money does not circulate, credit does. And no, money and credit are not the same thing. Banks do not lend the money that other people paid into the bank; it doesn't work, you would not have any money in your account anymore, and the bank would not have any money anymore either.

3 things that save the world: Money, money, money.

Why is our world so screwed up? Why can't we have clean energy, clean cities, great free universities for everyone, clean air, less war etc.? Because we keep voting for people who just make promises but couldn't deliver if their lives depended on it. And it always comes down to one last thing, the one thing that stops everyone from accomplishing what we all want.

Why can't my state build better schools? Why can't we have better educated and trained police officers? Why can't we all have a basic universal income and eradicate homelessness and poverty – literally in less than a few weeks? – forever! It is, they will tell you, because we do not have enough money to pay for it all. Then we have to ask ourselves, what is this *money* that they speak of? What is it and where does it come from?

Where does money come from?

In every government on earth, every representative, starting from city council members and PTA meetings on up.....if you can't explain where money comes from – you are not qualified to represent anyone, you can't speak of money, or budgets … you cannot put your name on a ballot, you cannot declare war on anyone, you cannot lead anyone to a better system, you cannot prevent hate, you cannot repair civility, you cannot educate our children, you cannot be allowed to speak of funding and deficits, you are not qualified.

Everybody should be able to answer this simple question: Given an initial deposit of $1K, with zero reserves, 100% lending capacity, over 5 steps, how much money do you create? That is an easy question. (don't stress, don't do the math, keep reading...)

The above question should be part of every high school finals. Fail that one question and you do not graduate.

We could easily afford to educate our entire population, with three times or 4 times as many teachers, in classrooms with less than 20 students. All leading to more harmony and understanding, which leads to a lot less crime, less hate, less anger, nicer streets, cleaner streets, and all that education and knowledge will lead to less pollution.

Clean up the oceans costs billions, so what? We could clean up the oceans, plant billions of trees and erect a few thousand wind generators next to super large solar panel farms.

We need to fix, build and rebuild thousands of bridges, roads, schools, dams, and we can do all of that, easily, we could do it in a heartbeat. You just need money to do it.

… if we just had the money to pay for it all, right?

I haven't lost you yet, have I? Keep reading, it's getting good ……stick with me here.

Why do we always appear to not have any money when we want it, but in a big nationwide emergency, we magically can borrow it? By the hundreds of billions? From whom?

"you're crazy, how we gonna pay for all that?"

How can we allow them to use the excuse that we are out of money, and that's why global pollution will kill us all? You see, we didn't have enough money to stop global pollution, sea levels were rising, cities got flooded, oceans got polluted, the air is polluted, we don't have enough farm land, we could not convert ocean water into drinking water, so mankind died of thirst… all of that, just because people can't answer a simple question? In other words, our children will all die – because we ran out of money? That is insane!

If money can solve so many issues, since we have made it so essential to our existence on this planet, why can't everyone tell you where it comes from? It should be as simple as the knowledge that the earth revolves around the sun, while the moon revolves around our planet, unless you really believe the earth is flat.

Not only do you need to ask: Where does money come from? You also need to ask: Why do my elected officials not know the answer to this? And why aren't they teaching us how this really works? How is anyone allowed to graduate from high school without knowing how this works?

'You have to earn money!' this is a really sad answer.

You can't earn money that doesn't exist. If there is not a bundle of cash on the table, then you work in the yard and dig a big hole, and an hour later there is a lot of cash on the table? Money does not come into existence just because you work. You can't earn something that isn't there. You can't earn money that doesn't exist. Where does money come from?

At this point you should be smiling, a bit. Its is silly that we have this thing, this money, ever since we were kids, and we never asked anyone where it comes from. Nobody has ever shown you the easy math and shown you how this stuff works. Now you will see how silly the numbers have become. Stay with me:

How did governments around the world, simply borrow hundreds of billions of dollars, and trillions over the last 50 years?? Where did that money come from?
(www.axios.com/world-total-debt-load)
The world's total debt, in early 2020 (before Covid-19), was over **$253 Trillion Dollars**. The world owes $253 trillion dollars. Yes, that number is astonishing. Look it up. Total government debts on earth have totaled over $253T before we added borrowed trillions for Covid-19 to it. Governments owe over $253Trillion Dollars. TO WHOM?

Did somebody have $250 Trillion Dollars lying around, in cash, just like that, and what was their plan? To just chill and wait for some people of earth to come by one day and ask to borrow some of it? Where did these $253 Trillion Dollars come from? You can't possibly think that some world bank "*earned*" and saved $250Trillion Dollars, do you? How would they earn that?

If they/we all borrowed $250 Trillion Dollars…don't we have to pay that back?

How do you pay back $250 Trillion Dollars if you don't have any money and you need to keep borrowing more and more? Why would anyone lend governments any more money, knowing that they can't pay it all back? And where does the money come from that they lent them? And WHO are "*they*"?

Here is the gist of what you need to know about our monetary system. The Federal Reserve Act was passed by the 63rd United States Congress and signed into law by President Woodrow Wilson on December 23, 1913. The law created the Federal Reserve System, the central banking system of the United States. It was called "federal" and "reserve", even though it is not a federal agency, and it does not have any reserves.

The idea was that all banks in our country work together, under the umbrella of a larger overseeing organization, that ensures that all banks in its system will apply the same accounting methods. In return for that stability, the umbrella organization, the Fed, will issue currency/money to the banks, so they can lend it out, and with that, we can run an economy. Simple, isn't it?

Let's say person A has a savings account with $1000. The bank that is holding that account, will now lend the $1000 to person B. Now person B has the $1000, and he/she will buy something with that borrowed money. Whatever they buy, the seller of those goods, person C, now has the $1000, and they will promptly make a deposit in their bank, for $1000.

Guess what, that bank that now has these $1000 from person C, they will turn around and lend that $1000 to one of their customers. Now person D has just received a loan for $1000 and they will go out and spend that money, which will end up in yet another bank and they will turn around and lend that money again to somebody else. A nice circle, right? That is the basic multiplier-effect in our banking system, worldwide.

You start with one deposit of $1000 and 5 banks later you have issued a total of $5000 in credits. Not money, just credit. The amount of money has not changed. We did not *add* any money in this scenario. We still only have the original $1000 Dollars from person A. None of the banks have any money left, they lent it all out. 5 people owe $1000 each in loans to their bank, 5 people borrowed $1000 each. But the amount of money has not changed. No new money was created.

Here is the problem. Every bank lent out the $1000 that someone else deposited in their bank. What are those banks going to do when their depositor customers come back and want to close their savings account? How is their bank going to give them back their $1000 deposit, if they don't have it anymore?

You see, banks cannot lend out the money that their customers paid in, because if the banks would lend you the money from other customers, then those other customers would no longer have any money in the bank, and the bank itself would not have any money either. On top of that, it would never create any new (more) money. We are not recycling money, we keep adding more and more money every day, printed, computer generated. And remember, you can't earn money that doesn't exist, you have to add more new money to the economy or there would not be any economy. This additional new money has to be printed. But it only gets printed, 'created', as debt. Someone has to borrow it, before it gets created.

So how does this stuff work?
Money is created out of thin air. Money does not come into existence because you work. Money is created as debt. Yes, debt. Every Dollar and every Euro that was ever created, was created as debt, and somebody somewhere has to pay interest for it all.

The basic premise is this: Our government is not allowed to just create its own money, otherwise they could just print a gazillion dollars and give all of us a few million bucks. No, they have to borrow it. In order to borrow it from the Federal Reserve, the Reserve Bank has to create the money, print it. I

The central banks will print/create the money that their government needs to borrow. They print it as debt, at the moment the government will borrow it, so they can ask the government to pay it back, at a later time. If they pay it back, then that process will eventually take that borrowed money out of the economy, thus pulling it back, which in turn is supposed to keep inflation at bay.

Governments have to provide some kind of collateral before they can borrow all that money. Well, that collateral is nothing more than an IOU, I owe you. A piece of paper, we write up some easy terms for it, you know, pay back over the next 10 years, pay 2% interest every year, and so on. What kind of IOU is that? It's called a government bond, or treasury notes. We print a government bond, then central banks, the Fed, print money, give that money to the government, they spend it on military war machines, and then we pay our government with taxes, then the government uses those taxes to pay the interest on the money they just borrowed.

That's it! (in a real easy nutshell)

You almost got it, take a little break if you like. We get right back on from a different point of view, to help you grasp this. It is really important that you understand this, so that you can see how silly it is that we can't save our planet.

The hard part is over. It gets easier now, read on.

CHAPTER 13

Ok, one more time. We can't just print money and give everybody on earth a few million dollars. It would collapse the whole world because if everybody had a few million dollars, it would all be worthless, since everybody has it. Somehow, we need to keep track and account for what we print, however, we can also print any amount we need, when we need it, for any project. Like emergencies, hurricanes, wars, Corona Virus Epidemics etc. That means, we can, yes, we can print any amount any time....

Printing it all will create inflation, no?

When we print the money we need, to pay for services that we receive, then we don't create any massive inflation. We can print 10billion dollars and manufacture wind generators, transport them, install them etc. without any state or government having to "borrow" the money, and then owing it back, plus interest. The 10 billion dollars did not create inflation, they created jobs and income.

Why do we have to pay back what never existed?

Our cities, states and federal governments print IOUs called municipal or government or treasury bonds. Yes, they print a Bond. It's just paper that says I owe you, and I will pay you back at a later time. That's it. Just a piece of paper. You could use a sticky note, but it won't look as good or official. But it is the exact same thing. Just paper. Simply printed. Or, nowadays just created on a computer.

Against that bond the Fed then prints actual dollars. Our 'central bank', the federal reserve, which is not a federal agency and has no money reserves, will now print an equal amount of US Dollars, against the bond. The bond in and of itself, that piece of paper, *that* is the reserve. The debt, the money owed, the contract, that sticky note, the bond, that is the *reserve* against which central banks print money. Nothing is backing up any money on earth, no gold, no silver, no moon rocks. All the money ever created, ever printed, is promised back, plus interest, against a promissory note. It gets printed because we 'promise' to pay it back, later, plus interest.

And the system works, it works just fine. The Accounting clearly shows how much is owed.

Are you still with me? You got it? You almost got it, that's good, it will get easier, believe me, keep reading. Now we are getting to the fun parts.

We need to print these papers, called bonds, in order to print these papers, called money, - then we need to print more papers called money, to pay back the printed money, so that we can cancel out the printed bond, to show that we no longer owe ourselves any papers... that is exactly how it works. And any 10year old can tell you that this is ridiculously insane.

Remember when this system was enacted and try to remember who the people were back then. Over 100 years ago these people had no interest in saving mankind, they were some of the most horrific racists on earth, their parents and grand parents had slaves, slaughtered native American Indians and ate squirrel for lunch. This was not some sophisticated society with free higher learning for all. Some people still thought the earth was flat. Snake oil was magic, and the sun was the center of the entire universe. It is, however, a main reason why we have so much pollution. It is the reason for the relentless pursuit of more profit and ever more growth.

The fact that all money on earth was created as debt, the fact that it has to be paid back, plus interest, that is the reason for inflation. When you print a $100 million dollar bond, then print $100 million dollars against that bond, well, then you have $200million dollars of values in the economy. That is inflation.

There, now you know, we created money. We decided to print it, as debt, so we will not come up with the crazy idea of just printing too much of it. We have what is called 'checks and balances' to avoid run-away inflation or a collapse of our system. But that's not quite true. Just before the attack on the World Trade Center, the Pentagon was in a bit of a jam, they had to admit that they *lost* $2Trillion dollars. Yes, simply lost it. Just like that, poof, it was gone, they said they had no idea where it went. Rumsfeld was supposed to provide some congressional hearing with an explanation, but that never happened. A plane flew into the Pentagon, right into the part of the building where their accounting offices were. It was all gone, and so were $2Trillion Dollars.

However, we did not have any kind of inflation because of that. Interesting.

Just a few more things and you will see all the connections. Can you go on or do you want to take a 5 min break? OK, come back after, this will explain a lot right here......

What is a write off? In our agreed upon accounting methods, we say that any amount, or credit/loan that cannot be returned, can be *written off*. What is this actual 'writing'? Well, in the old days, we did not have computers, yeah, I know, you knew that already. Everything was handwritten and later typed. The actual process of a write-off was then a ruler, placed over the amount of money that would be written off, and a pen was used to then cross out the amount of money. Literally, "crossed out" a dollar amount, wrote a little number next to it. Then there was a ledger that had a list of the notes, to correspond to a number, that then had a written explanation why the amount was no longer collectible. It was ~~written off~~.

It is that easy. Put a line through it, it is gone. No longer relevant.

How do banks lend money? The simplified system for an easier understanding is based on a 10% reserve. Meaning that a deposit in the bank of $1000 is not to be lend out in its entirety, you should only lend out $900 of that. That would give the bank a 10% ($100) reserve.

When you start with $1000 on the multiplier-effect in the banking system, but now with a 10% reserve requirement, you end up dwindling that number down quickly. You receive $1000, you can only lend out $900. The next bank receives $900, but they can only lend out $810, the next bank gets $810, but they can only lend out $729 dollars and so on. There is an end to this supply of credit. Everybody has to hold back 10% of the deposit that they receive, then it will take 10 steps and the money is gone. You cannot endlessly lend out the same amount of money all the time. That is the easy answer to the final high school exam question.

But then the Fed requires reserves for money that the Fed is willing to give to the banks in their system. Say 10%. Alright, how about you take the entire $1000 deposit from your customer, and designate it as *'reserves'* for the Fed. Now follow the money here. The Fed will give your bank $10,000 dollars to play with. After all, you do have 10% reserves for the Fed. You have $1000, from your bank customer, that is your reserve. That reserve, those $1000, that is equal to 10% of $10,000. Ergo, your bank has reserves of $1000, therefore your central bank is willing to lend your bank an amount of $10,000, at the prime rate.

Many of you think that just because it is called 'reserve' it must be something that the bank owns or earned. Not so. It was never like that. However, the name given is precisely used to confuse you and make you think there is more security in the system, or that banks are careful because they have their own money at stake. They do not, they never did. That is why they are so reckless.

It was borrowed, the bank borrowed the money from the Fed, and all the Fed wanted to know is if the bank has reserves. The bank uses your deposits as reserves. They also use your mortgages, government bonds, stocks and anything else they can get their hands on that the Fed System allows and accepts as 'value', against which the bank can then borrow money from the central bank, at the prime rate of just 0.25%, while you pay your bank 5.25% interest.

Now I am done torturing you. No, really, I mean it. Hopefully, you did not nod off or jump out the window. It really is not that hard to understand. WE *print* money. (where else would money come from?) WE created our monetary system. That means that we can create our own new system that will help us to save our planet. Why would we not?

Well then, how is this money part helping us with global pollution? Easy, we can create a new system that is only for environmental purposes. The Green Dollar, the Terra Dollar, the Global Coin, or whatever you want to call it. We do have crypto currencies, we do have smart contracts and we can issue new money without debt, just to help ourselves save our own planet. No, really, we can do just that.

Nobody could possibly be against saving our planet, could they?

Let's call it our Global Environmental Credits, GEC. Or, please, come up with a better name, somebody, please figure out a new name and a nice shiny logo for our new money. Until you send me a new and better name, let's call it **GEC** for this book, for now.

What if we create a new monetary and accounting system, that runs *parallel* to our Fed system? We are not abolishing the Fed, not yet.

We can issue new money through smart contracts that allow us to trace every amount to ensure there is no fraud in the system, although there still will be. But none of that will compare to the $2Trillion our Pentagon conveniently *lost*.

Carbon offsetting programs such as the one implemented by California have the potential to generate revenues and encourage innovation. Critics, however, have suggested it has a number of design issues.

One such issue is the fact that California's carbon credits do not expire. This could allow companies to stockpile credits and ignore future cuts to the emissions cap. Another concern is that the companies covered by California's cap and trade will simply pass their higher costs to the consumers. (visualcapitalist.com)

They do not expire? Why not? For crying out loud, even my car registration expires, even though it is not a fruit and my car didn't run away. Toxins don't expire either, then why on earth is this program not requiring companies and operators to renew their carbon offset / carbon credits every year? Are you for real?!

Who took a bribe here to make sure that polluters and toxin terrorists only pay once and then never again? Can we please find some leaders with the balls to make some real impactful changes?

Other inefficiencies within the program may exist, but its benefits are hard to ignore. By the end of 2019, the revenue generated from California's carbon credit auctions totaled $12.5 billion. Of this amount, over $5 billion has been invested in GHG reduction projects to date.

You see? They can, yes, they can do something, but we are still not willing to go the next steps. Why is that? What are you waiting for? This is just a prove of concept project, this cannot possibly be the whole program, right? And how many States have this program? Only 12. Yes, only 12.

Where is our national policy on toxic pollution? What, we don't have one, we still don't have one? That's because you elected politicians that can't get it done, people without a plan or paddle. People bending over for the polluters.

It is still far better, right now, to simply issue a new form of payment to our own economy, GEC debt free money, and create all the jobs we need at home and clean up our toxic mess, while building our energy of the future, instead of waiting for something to change by itself, maybe, in the next ten years or so, or not, or never.

This new money, our GEC, can also provide the UBI, the Universal Basic Income that we will make available to all the people that lost their jobs when we shut down those disgusting coal mines... or so I hope I will be able to say to my son in our not so distant future.

We can have new money, without debt?

CHAPTER 14

Yes, no, wait, what? Money? We print it? Yes.
Then we owe it back? Yes.
To ourselves? Basically, yes.
Plus interest? Yes.
We don't own our central banks? No, never have.
Our governments have to 'borrow' money? Yes

Who came up with that system? Your great, great grandparents, around the turn of the 19th century, then it was voted on by congress, in 1913. Although, none of the members of congress had fully understood nor read the act they were going to bestow on the rest of our planet, nor did they comprehend the multiplier effect or the meaning of reserve requirements. We are working from a system invented by greedy racists with questionable understanding of the universe, no concept of global warming, who thought nothing about dumping toxins in a river and beating their wives for fun, yeah, that was legal back then too. We had barely just invented the radio! Our *'modern'* money system is over 100 years old. Good grief.

Money is printed out of thin air. Money doesn't circulate, credit does. Banks don't lend out the money that you paid in; they use your money as reserves to borrow money from their central bank. Financial fiascos occur when banks no longer have any reserves against which they can borrow from their central bank. During the 2008/9 financial crisis banks didn't run out of money. They just didn't have the reserves to back up all the money that they had *already borrowed* from their central banks.

All those 'assets' against which the banks borrowed from the Fed suddenly didn't have any value anymore. Remember all those mortgage-backed securities? Well, they were not really that secure, they were extremely volatile, and banks used them as 'value', as 'assets', as 'reserves' against which the central bank then lent money to the banks. Those securities became almost worthless overnight. Oops. Then our governments decided to issue even more government bonds, as collateral, to borrow more money from the same central bank, from the Fed, to then give that money to the banks, so the bank's accounting could show real cash, which then appeased the central bank. Yes, it is exactly that stupid

And most banks are still receiving *subsidies* from our taxes, even today! When did your *officials* explain that to you? When did your teachers tell you how this works? They never did, did they?

CNN.com, November 17, 2020

Governments and central banks have promised to shell out **$19.5 trillion** since the coronavirus erupted to "put a floor under the world economy," according to the International Monetary Fund. Some countries need even more help to recover from the crisis, but they might not get it.

Yet despite the unprecedented scale and speed of the rescue, which cut taxes, paid wages, granted loans to small businesses and took interest rates to record lows, the global economy is suffering the worst recession since the Great Depression. Economic activity and employment in large parts of the world, including in the United States and Europe, remain well below levels seen before the pandemic hit.

Asia's economies are bouncing back. The West is headed in the other direction.

This is, all of this, this is newly freshly printed money that didn't exist until they just now created it. Why are so many in uproar over these trillions then? Because we have to pay it all back, plus interest.

All of that, all those trillions, are money that never existed before. It was freshly printed. As debt. Somebody has to pay back all that money, that somebody is me, with my little job, with my little income, I will have to pay more taxes to my government so they can pay back what they borrowed. Instead of taking a ruler and just ~~writing it off~~ at the end.

I think you can see by now that this system is not going to help us save our planet. You cannot use the system that got you in trouble, to get you out of trouble. The very definition of insanity is to do the same thing over and over again and expecting a different result.

The idea is to issue a new kind of money, not as debt, but as needed to fund the construction of desalination plants, wind farms, wave energy projects and the likes, on a grand scale, all at once, all over our country. And we will agree on a new accounting system for the issuance of this new money to pay for it all, including a UBI. By the way, a universal basic income is not something that will have you live lavishly and never work again. It will be just enough so that 3-4 people can afford to live in a one-bedroom apartment and pay for their basic needs. Remember, we do not need coal mining jobs, we need income. Let's give them peace of mind, unemployment money and UBI, which should also pay for their healthcare, while we retrain those that want to get a new job. We will train them on how to build, manufacture and maintain wind generators, wave energy platforms, bug farms and how to utilize the sensors for new EPA agents and how to write a report on the polluters.

Yes, we can print, or digitally create all the funds we need to pay for and build all the things we need to help us save our planet. It will not add to any inflation, and at the end of it all, we can just write if off, or ~~convert~~ it back to whatever currency and monetary values we will have in 20 years from now. And even if it adds to inflation, so what? As long as it helps us save our planet, who cares? Inflation does not kill our children, pollution does.

Let's try a parallel system, for 20years. Not debt money, funds issued only for environmental projects. You can't use it to build tanks. Do you remember how much it costs to build a city? Well, here you go, voila, use the newly created GEC funds to build it. You can start that project tomorrow!

Need more funds to expand your offshore windfarm? No problem.

Need more funds to install more solar panels for that development? No problem.

Need to plant 10 thousand hemp seeds to cover the barren hills from the last forest fire? No problem.

Need more funds to clean up the riverbanks from the coal ash pollution? No problem.

Need more funds to grow plankton and biodiesel fuel to help the fish find food? No problem.

Need more funds to expand the bug farm facility? No problem.

Need more funds to build more clean energy trucks to haul the salt from the desalination plants to a dried-out salt bed? No problem.

It costs about $80million to build a desalination plant that produces 10million gallons of fresh water per day, MGD. $250million for a 35MGD and $650million for a 100MGD plant. Cool. Let's build a hundred of these on the west coast, another hundred on the east coast, and let's put a few on higher grounds near Texas. All that fresh water we create can be pumped right into our almost dried-out freshwater aquifers. Let's fill them up, all the way to the top. Our farmers need that water. All we need to do is built them, create the energy they will need to run, like windfarms, wave energy, solar panels etc. How many thousands of jobs will that create? How much will that help us over the next 30 years while we are facing ever-increasing temperatures and droughts?

We don't have to fear droughts anymore, we have all the water our agriculture farms require, and then some. Need more water? Build more of these desalination plants. Can't afford to build them? Here, have some more GEC money and get started.

How do we pay for all of that? We create as much GEC money as is needed.

Oh, yeah, all those newly created funds are issued without debt, they won't have to be paid back by our government or our taxes, that might actually help us and we might no longer die of hunger in 50years from now. That could be helpful, eh?

Maybe you are not quite there yet to understand a UBI. You may have never heard of it, seen it, or researched it. Most of your 'officials' haven't either. But we have them, we've had them for a long time. Here is a wonderful graphic that shows you all the UBI programs that have been used, just none of them have ever been used on a grand or global scale.

https://www.visualcapitalist.com/map-basic-income-experiments-world/

CNN, November 2020, Christine Jardine, a Scottish politician who represents Edinburgh in the UK parliament, was not a fan of universal basic income before the pandemic hit.

"It was regarded in some quarters as a kind of socialist idea," said Jardine, a member of the centrist Liberal Democrats party.

But not long after the government shut schools, shops, restaurants and pubs in March with little warning, she started to reconsider her position.

"Covid-19 has been [a] game changer," Jardine said. "It has meant that we've seen the suggestion of a universal basic income in a completely different light." In her view, the idea — sending cash regularly to all residents, no strings attached — now looks more "pragmatic" than outlandish.

Experts see the coronavirus pandemic as a world-changing event that could result in a similar tectonic shift. Job insecurity caused by the pandemic, however, appears to have generated new levels of support for the policy. One study conducted by Oxford University in March found that 71% of Europeans now favor the introduction of a universal basic income.

"For an idea that has often been dismissed as wildly unrealistic and utopian, this is a remarkable figure," researchers Timothy Garton Ash and Antonia Zimmermann wrote in their report.

It probably helps that the pandemic has helped normalize cash transfers from the government, said Nettle, who has also conducted his own polling. According to data compiled by economists at UBS, nearly 39 million people in the United Kingdom, Germany, France, Spain and Italy were being paid by governments to work part time, or not at all, as of early May.

And how do you think they ended that report? Here it is: Critics also raise fears about the broader economic ramifications of such policies. Some worry, for example, that providing a universal basic income could lead to a spike in inflation.

One more thing on inflation as a counter argument. We cannot afford to create a few trillion dollars of GEC money; it will create inflation. Yeah? So what? Let it. No, seriously, if inflation is the big bad wolf, bring it on. AT LEAST WE LIVE!! Who cares about some inflation in 30 years from now? What are they gonna say? Oh, yeah, they all died, we had no trees left, we poisoned all the water, we had no food and it was 120 degrees all over the planet, but, hey, at least they died without inflation. Are you kidding me?

What is more dangerous? Inflation of a few percentage points, maybe in 20 or 30 years from now, or, or is it more dangerous to have $26Trillion Dollars in debt that we will never be able to pay off while we borrow more every day? And on top of that debt we still have all these toxins in our environment, and we keep adding more all the time, but god forbid we might have a bit of inflation after we save our planet and our children with non-debt GEC money?

Not only will this GEC money pay for cleaning up the mess, but it also builds our clean energy future, while providing a couple million new jobs, all at the same time. And you want to start yet another useless committee and waste our time with endless debates about inflation? Are you high?

You have to be able to see that anyone telling you about *inflation* as a reason not to save ourselves, is a corrupt piece of crap working for the toxin terrorists, right? Why else would you possibly be against creating our own solution for our own problems without having to pay for it with exorbitant interest and be beholden to an imaginary bank? They are either corrupt or stupid. Well, probably both, but they also heavily count on your ignorance and lack of understanding how this money thing works. They know that most people do not understand where money comes from.

When Elon Musk creates colonies on Mars, do you think they worry about inflation on Mars? Which is more important, saving the planet or avoiding inflation? And why is there even a discussion about that?

Besides, we can pull that money back out of the economy. There was this idea of taxing only the rich. The billionaires tax.

That doesn't work. At that level, they do not have any income, nothing that you can tax, unless you believe you have jurisdiction in Panama or the Cayman Islands. The only way to take money out of the economy, and putting it in the cities that need it, is to add a **VAT**. Value Added Tax. It basically adds a tax on high priced items. Say you buy a $300,000 Ferrari, now you pay 10% VAT, that means you pay an extra $30,000. The car will cost you $330,000. You buy a house for $2mil, now you pay $2.2mil. You want to buy a private jet for $40mil, now it will cost you $44mil. And you know what, it works. And people that buy a $5mil boat don't really care if they have to pay $5.5mil for the boat. It does not make things more expensive for the average taxpayer.

Another insanity is the States that want companies to *pay* them $400 million just to create a windfarm offshore, they want $400 million as a license to allow them to build a windfarm that provides clean energy to us all. Really? Why?

That means they need to sell energy at a higher price, for profit, and to make back the stupid $400 million. Why would you want to deliberately make our renewable energy projects more expensive? Maybe so coal will keep looking cheap?

States are so shockingly poor because of their mismanagement and bond issues, they are so desperate for money they will charge a half billion dollars as a "license", which makes the whole darn thing far too expensive... it's upside down.

We need to train and certify an extra 5000 accountants and managers, at least that many to start, that will be imbedded with every GEC project and company working on it. They will make sure that the forensic accounting can show where the GEC money went and make sure it is spent only on approved projects to help get things done and to prevent fraud etc. No company shall have more than 5-7% profit margins on these projects.

Stop talking about making money and profits when we need to save our planet. You won't be able to bribe the GEC managers either, they won't be on your project long enough before they go to another. We will rotate them. You would have to bribe all of them in order to get away with something shady. Remember, they do not work for the companies, they work for GEC, they work for us.

We will rotate these GEC Auditors and GEC Managers from one project to another, they will receive a modest reasonable salary, not stock options. There will be no stocks, no publicly traded anything, nobody will be able to 'invest' in GEC bonds, nobody will pay back this money, nobody will profit beyond a reasonable margin of 5-7% and those involved will then later be able to convert that GEC to other money, as we write off these credits.

You can go and spend your GEC salary just like you spend any Dollar, it shall be accepted everywhere, any store that doesn't want to sell you an ice cream cone for GEC credits? Fine, no problem. We will close that store and take the owner to jail. That message will get out real fast. This is the money we use to save our lives AND yours! You better accept GEC as legal tender, or you will lose all your assets... get it? Got it? Good. Now let me have that mango flavor over there and put some sprinkles on it, then let's get to work!

It will be the job of the GEC accountants and managers to get this all done and break through the procrastination and all waived red tape on these projects. GEC projects shall be immune to red tape, they shall have absolutely priority.

"Why on earth has that truck not left the yard yet, it was ready 20 minutes ago...!??" asked Keisha in her usual demanding tone.

"Well, there was an issue with some of the tags...."

"Oh, hell NO, there was an issue with whaaat? And NOW I hear about this for the first time...what da hell is going on here... is this how you run this outfit ??" .... She turned around and made sure that everybody could hear her now....

"LISTEN UP EVERYBODY this is NOT how I manage things around here......get your butts in gear and make this happen, I want to see another 8 trucks get out today, and I am NOT taking any sissy excuses from anyone is that clear?!! We need this material out and delivered, that windfarm is NOT gonna run without these omni blades, and we will NOT be the reason for ANY delays, is that understood??!!!" they looked at her with cringeworthy expressions of anxiety... although the other GEC manager was not much more lenient. But then again, the other manager admitted it was her first time as a GEC manager, she had just come out of a school to be retrained from her old oil logistics job. She did great, and she didn't raise her voice quite as much....

"Are you sleeping?!" She looked at him as if she were burning her laser beam eyes right through his brain. " I can see you thinking something, but I can't see you working ... why is that? Are you dreaming? I'm trying to save a planet here, we are running out of time, and you're just standing there staring a hole in the sky, I'd appreciate it if you could get back to work !!"

They all rushed back to work, falling over each other, started shouting orders, it was as if they suddenly found themselves operating a nuclear submarine with the enemy bearing down on them....

- (PINGING SOUNDS, PING, PING, PING)
- Hold up.
What do you hear?
(TINGING)
HOLY COW, he yells, Torpedo in the water, TORPEDO IN THE WATER!!
bearing 3-3-5 at 800 yards!
Left full rudder!
Make your depth 500 feet!
- put her on the bottom in that canyon
DIVE DIVE DIVE
Maximum dive, full down angle, hold on to something
- Six hundred yards!
Where did this thing come from?
- All ahead flank!
- Captain, there's no time.
- Four hundred yards!
- hurry up!
Countermeasures, full spread
Full spread aye Sir
- Terminal homing!
PING PING PING PING
- captain!?
Get in that canyon!
NO TIME SIR
Brace for impact
(beeping and swishing sounds from the decoys)
100 yards!

BOOM. The sub was just sliding over the edge of the ridge, the torpedo hit the ground behind it, as the sub was gliding at full speed into the depth of the canyon, shaking from the impact and pressure wave of the explosion behind it.

Meet Keisha, nickname: Boom!

Yeah, we need a few thousand of these managers to kick ass and take names and make sure we can save our planet. Get on it. I know a fabulous woman from Jamaica that has just the right kind of attitude and voice for this, oh boy, she will shout out a million commands a minute. You better not miss a word, or you will get crucified… more than once… yes, she is an absolute sweetheart. (I know you're reading this bae…☺)

I promise, no more math questions in this book. Read on, it's all coming together now.

CHAPTER 15

Why do you think Elon Musk wants to get off this planet and go to Mars? Don't you think he can read the same numbers and reports I read? Of course, he knows. Plenty of people know the trouble we're in, but somehow, we can't find the right leaders to get us out of it, not yet anyway. And somehow that has something to do with money.

When I said earlier that *growth* is the destruction of mankind, I meant that. I am not asking to reduce our economies; I am asking to reduce the damage that all that economic growth is causing – those are two different things. Stop celebrating profits, unless you earned them with an absolutely clean operation that didn't cause any harmful emissions.

Unless you can show that you created zero toxic emissions, you should not be allowed to pay bonuses, or profit sharing or dividends. You should not celebrate how much money you made if you didn't pay to clean up your mess. You actually did not make any profit at all if you just passed-on the costs for cleaning up your mess. You need to pay for your mess before you say you had no money to install filters, buy carbon credits, or pay Toxin Taxes, but you were able to pay bonuses.

Stop celebrating growth. Start celebrating and promoting the companies and people that reduced their pollution. Don't invite any CEO on TV that is running a company that cannot clean up their mess. They do not deserve to promote their stock for free. You're culpable if you help them promote their stocks when you know that they do not clean up their mess and that their toxins are killing us.

How about Nasdaq and the Dow Jones and DAX or any other trading platform delist companies that cannot clean up their messes? Your stock price will be frozen, trading will come to a screeching halt, unless you can pay to clean up your mess. That would change some boardroom dynamics really quick. I bet they would all find a way to at least pretend to care.

It is not one problem that we have, it is all the problems that we have that worry me. Remember, you cannot swim in 50% of our lakes. 50%. At what percentage point did you want to make changes? Do you really want to wait for that number to hit 75%? Really? 90%?

And why on earth, for all that might be holy and exist in all the parallel dimensions, why did you not get really mad when that number hit 20%? You waited some more, you did nothing, you just sat there and watched them dump their toxins in our lakes. Then, at 30% destruction, you said, ah, screw it, destroy some more, see if I care!? Say what?

Do you not understand that if 50% of all of our lakes are no longer save to swim in, that means that the other 50% of our lakes are already being used as toxin dumps? They are toxic, they are being destroyed, they are just not so toxic, yet, that we would say don't swim there. They are toxic already, all of them, just 50% of them are not quite too toxic, yet. And apparently you want to wait until that number reaches what percentage? How long do you want to wait before you begin to see that our current system is not working? Our current leaders are not making any progress, they only make excuses. And they still use words like 'money' and 'inflation' to confuse you and make you belief that it isn't their fault.

Can you see it? Do you at least begin to see that WE are in our own way to build the things that WE need in order for US to live on OUR planet? Here is that sentence one more time. We are in our own way, to build the things that we need, in order for us to live. Let that sink in, slowly. Although it makes perfect sense, it is also very stupid to have to admit that we did this to ourselves. If you were to try to explain this ridiculous predicament to a class of ten-year old's, they would look at you like the crazy person that you are.

"......yeah, uh, why did you adults do that to us?"

"Well, you see, you know, it's kinda like, you know, we didn't have the money to help rescue ourselves......and then...."

"Where does money come from?"

"Uh, well, you know, it's complicated, and you are just ten years old, so I don't think you would understand, if I tried to explain and...."

"Don't adults just print money?"

"Well, Timmy, yes, actually, we kinda do, did your mommy tell you to ask these questions......?"

"Then why don't you just print more to save our planet?!"

"Well, you see, you must understand, it's complicated, and, you know, we, well, we will have to pay that back, and banks want to control it, and if we print too much, then......."

"Pay it back? Why?"

"It's complicated, but we have to pay it back some time, right, and it costs......."

"My mom said you are an idiot. We can always print more, and we can just draw a line through the past. Why don't you do that?"

Bravo, Timmy. Class dismissed. And say hi to your mom for me.

Where are all these excuses leading us? Well, to be absolutely blunt about it, they lead us all to death. And we will end up there much faster than anyone seems to be telling us right now.

In 2008, an informal survey of experts on different global catastrophic risks at the Global Catastrophic Risk Conference at the University of Oxford suggested a 19% chance of human extinction by the year 2100. And now we are in 2021 and it has gotten a whole lot worse since then. If I am saying maybe 50years, that would put us in 2071. We are basically just 30years apart on this. They said that in 2008, before Global Warming got as bad as it is today.

Humanity will kill us a lot quicker than pollution. And here is why:

"*I say, I say, boy, I say, stop digging a hole...!*"

Obviously, you cannot ask any of our current political "*leaders*" to get involved in this, they don't even know where money comes from, or they won't admit to you how it works. Nor are they willing to look at anything that will or might disrupt their status quo. As long as our current leaders have no clue how bad our situation is or refuse to admit how dire our planet's eco systems are, or what we can do to help ourselves, there is no point in talking to them. We need a whole new breed of leaders, teachers and politicians.

Scary question: How long will it take to get those new leaders?
Where are they? How do we find them? How do we teach them? How do we test them to prove that they truly understand these issues and they will actually vote in congress to do something about it, and do it now?

The rate of ice sheet losses, the rate of temperature increases, the rise of sea levels, the ever increasing amounts of pollution, the growing number of people on our planet, can you really add it all up? Well, I haven't seen that all-encompassing prediction model yet. And here is the problem. There is no model that showed Brazilian wetlands on fire, there is no model that showed the arctic on fire either. There is no model that shows you that more droughts lead to more terrorism, which leads to more local wars, which lead to more regional wars. There was no model that warned us about the tundra exploding.

Due to wars we have even less food in the global export market, which leads to more starvation, which leads to more refugees, which leads to more regional economic systems to collapse, which leads to more pollution. There is nobody left to clean up the cities and refineries and gas stations that they left behind. They will be rotting away, burst eventually and just add to our toxic misery.

Will you really clean up, I do mean CLEAN up Miami, or Louisiana or Bangladesh, all of it, before the waters rise too high and swallow it all? Will you really clean out all the oil storage tanks, all the gas stations, all the factories, all the garbage, you will go and take it all out, so it won't pollute our oceans any further once the water level has risen so high that you can't live there anymore? Really, you will do all of that? Cool. Thank you so much.

I'm just curious: Where are you going to put all that toxic and hazardous waste before the city goes down under? Will you ship all that crap to Malaysia too? How will you pay for the costs of cleaning up Miami before you cover the costs of a few million Floridian refugees needing a new city? Where are you growing all the crops to make the food that all our refugees will need when they arrive in these newly built cities? Or are you really going to shoot all the refugees trying escape from Florida or Louisiana? Or how about we just lock them up in camps and cages?

The connections will kill us a lot sooner than most people can fathom at this time. When our destruction reaches a point where most people cannot see a way out anymore, then the rest of the way will accelerate at incredible speeds. It's the compounding effect that is missing from all prediction models.

Let's try this. You go sky diving. Yeah. Fun. Only, you did not bring your parachute. You jump out at 9000 feet, you smile, you feel great. You know that the ground is way down there, far away, so you're not worried. You're having fun, spreading your arms, taking in the scenery, and you can fly. At about 7000 feet you are still having fun. How about 5000 feet? Can you make out buildings on the ground? Can you see the highway nearby? Can you see cars on the roads? Getting close, aren't you?

Are you still having fun at that time or do you have that nagging thought in the back of your head that you might not survive your impact, without a parachute? When will you begin to change your behavior? When will you start screaming? Are you trying to tell me that you will only feel anxiety and scream the last 50 feet before impact, or will that feeling have set in just about at 3000 feet?

Same question here: When people can see that we are not making changes in our monetary system, or creating one that can save us right now, or that we are not making any meaningful changes in our toxin policies, and we just keep on polluting and killing ourselves, at what point will they start to change their behavior? How long will that take?

With *behavior* I do mean paying their mortgages, taking kids to school, stopping at a red light, paying their bills, not steal anything, not shoot anyone, etc. You know, civility, social interactions with other human beings in a pre-agreed upon circle of life and behaviors. When do you think society will start to fall apart? Where is the prediction model for that?

When a few million refugees try to leave Florida, Texas, Arizona, Louisiana and the likes, and they all try to move to higher ground, those neighboring States must have the cities for these people ready to go. If they do not have those cities built by the time our refugees will arrive, they will be dealing with a lot of riots.

And those won't be peaceful and joyful parties in the streets, those will be people shooting each other over a gallon of water in a Walmart parking lot. When will the military get involved? How many riots, shootings and dead bodies do you want to see in your streets before you call the military, and when will they start shooting at their own brothers and sisters? How much time, how many years before the last day on earth will it be - before we start to fall apart as humans? What do our prediction models say about that?

Let's say we have 100 years left, 100 years until our planet is so hot and so polluted that it will be impossible to even find fresh clean drinking water. OK, so that is 100 years from now. Do you think society will stick together until 99 years from now? Will we just all behave for the next 99 years, knowing that the next year *after* that will be our last? We will only fall apart in the last 1 year before it all ends? Is that what you're thinking? Really?

Will people and societies start to misbehave 20years before it is all over? Will you still pay a mortgage if you know that in 25years from then we all die?

Will you even be able to get a mortgage in the last 30years of our planet's life? Really? What bank is going to lend anyone any money, when they know that all their customers will be dead in 30 years? And the bank itself knows it will be dead in 30years. Are you trying to tell me that the last 30years will be as civil and friendly, wrapped in harmony and peace, as the years that lead up to the total destruction of mankind? You think so?

Do you think anyone will get excited about the latest Nike shoes, when we all know that our planet is dead 30years later? When we can see and watch the water wars on TV, when we can see the slaughter of a few thousand demonstrators in Eastern Europe, mowed down by heavy machine guns from their own government, because they voiced their opinions on the latest food rations, that is the point it falls apart? Of course not. It will happen a lot sooner than that and you know it. Ergo, our prediction models are just based on wishful and unsubstantiated hope, our prediction models left out the human factor. Our current prediction models have a huge problem with confirmation bias and sheer blind eyes that overlook and totally ignore humanity.

Who is collecting pensions or IRA payouts or what government has funds to distribute social security checks in the last 30years of your 100year model? What wealthy neighborhoods will keep you save when hundreds of hungry and angry people start climbing the walls to your mansion? What more if they know that you run some kind of company that is polluting their lakes? Do you really believe all that money and all that private security will keep the mob out of your living room? How many bankers, CEOs and members of congress will likely get shot and killed by angry mobs of people that can't get any clean drinking water or get their kid to a hospital because it is throwing up blood?

If we say that we have 100years left, that means maybe 70years, because the last 30 out of those 100 will not be pleasant, not at all. Now if I say that we have maybe 50years left, because I can see too much garbage, too many toxins, 3billion+ more people, billions more in resource losses etc... then those 50 years begin to look really scary in about 30years from now and I know that we, our society, will fall apart in about 40 years from now. That means we all have about 50 years left in total, unless you implement drastic changes, and you implement them now.

In what year of our toxic destruction will people begin to shoot and kill company executives, senators, mayors, city council members, bankers, etc.? In what year, at what point of global warming damage or pollution caused destruction will there be enough momentum to tip entire groups of our own people to a point of '*nothing left to lose*'? Which prediction model shows me the time frame for 'nothing left to lose'?

Can you really fathom what that means? Have you ever been in a position of 'nothing left to lose'? Do you have any idea what that means to your mind when it doesn't matter anymore if you live or die, right here, right now, or if you take anyone else with you? Nothing left to lose is an incredibly powerful human emotion. It is exceptionally destructive. Take another look at the refugee boat in the Mediterranean, take a look at the caravans of thousands of people trying to escape their broken countries. Those are the faces of nothing left to lose. But they are still hopeful, they still have a small sparkle of life left in them. Do you really want to see what can happen when that sparkle is gone? Without that sparkle, the capacity to destroy our lives is extraordinary.

What prediction model truly accommodates its algorithm with a human component of total desperation, and nothing left to lose? None. It can't be done. But we have to think about it, you cannot ignore it anymore. If you can't see this happening, then you're about as sharp as a bowling ball.

There is another question that comes out of this volatile dangerous human factor. When we know that we only have x amount of years left, and our societies begin to fall apart, what is going to prevent some rogue military leader from launching their nukes? After all, he watched both of his kids die of stomach cancer, caused by some dirty water that the kids drank when they were out playing at some lake.

His wife was murdered last year during a supermarket rampage, and the only reason he still has half-way decent food is because he is in the military. That seems to be the only good job one can get in these dying times. But he also knows that America did the most damage to our planet, America refused to lead the world out of the mess it created, and they just didn't care about any other country or even their own.... maybe this is his big chance for payback. Maybe he can finally get some satisfaction by shooting at America.

Why should he not fire the nukes under his command? What difference will it make if we all know at that time that we only have a few more years or even a decade left? Why should he not fire them all? What difference does it make to him? Why should he care? What if he has nothing left to lose?

How much longer do you want to ignore the human factor in our global pollution problems?

If we can't see any real, massive, huge impact and life changing shifts in our pollution policies, over the next 15-20years, our societies will start to fall apart. (and the Pentagon knows this). That failure, that imminent collapse, will then speed up the destruction of the rest of our systems, thus it will dramatically shorten the time we could have left, that puts us at a point of no return, that puts us all in the category of NOTHING. LEFT. TO. LOSE.

That tipping point, the point of no return, is approaching ever faster and we are speeding it up with every year that we don't implement drastic changes to stop polluting our own planet, or your own State. We act as if we want to get to that point, as fast as possible. Are you suicidal?

Dr. Paul Ehrlich, Stanford University: "We are the only species.... determined to destroy ourselves."

Here is a flash back to what I said about coal and toxin terrorists.... yes, yes, they know exactly how not to pollute, but you make it so easy for them, you are practically begging them to poison our children. Here is something from propublica.org: *Georgia Power paid top dollar to buy land from residents living near waste sites at its power plants. Environmentalists fear it's a tactic to forestall the cleanup bill from new regulations for coal ash. Over the past several years, utility giant Georgia Power has embarked on an unusual buying spree, paying top dollar for people's property in places where cheap land was easy to find.*

In 2016, it bought a veterinarian's 5-acre lot in the rolling hills of northwest Georgia for roughly double the appraised value. The following year, it acquired 28 acres of flood-prone land in southwest Georgia's pecan belt for nearly *four times* what the local tax assessor said it was worth. By the year after that, it had paid millions of dollars above the appraised value for hundreds of acres near a winding gravel road in a central Georgia town with no water lines and spotty cellphone service.

Two things united the properties: They were all near coal-fired power plants that generated toxic waste stored in unlined ponds at those sites. And they were all purchased after the Environmental Protection Agency finalized new regulations in 2014 governing the disposal of such waste, known as coal ash. All told, the utility paid over $15 million for nearly 1,900 acres close to five of its 12 power plant sites, according to an investigation by Georgia Health News and ProPublica.

The costly land purchases offer an enormous potential payoff to Georgia Power. They may allow the utility to forestall millions of dollars in cleanup costs outlined by the December 2014 regulations.

The Atlanta-based company is trying to convince regulators to allow it to leave more than half of its coal ash — around 48 million tons — in unlined ponds at plant sites spread across the state. Environmentalists believe the safest way to dispose of coal ash is to move it from unlined ponds into landfills that have a protective, and more costly, liner to prevent contaminants from seeping into groundwater — the source of drinking water for people who depend upon wells.

They have the money to buy all that land, they also have the money to clean up after themselves, they have the money to prevent toxins from getting into our lakes and drinking water, BUT IT IS CHEAPER to keep polluting!

You did that. You voted for the people that create the laws that allow these criminal enterprises to continue to poison us all. You killed your own kids.

What can you do? Let me show you.

CHAPTER 16

What can YOU do? Spread this book around, make sure people read, listen, discuss, provoke, insist, don't let up, be relentless.

How about you get on all your social media accounts and start pushing a few hashtags #GlobalWarmingTruth, #StopKillingUs, #TorchedNScrewed, #NoPlanNoVote, #ToxinTerrorists, #climatechangefacts......

It's not getting better on its own. It's not gonna stop, nobody is coming to rescue us. Nobody is going to invent that one thing that changes everything. There is no one thing, we need all possible solutions to work all at the same time. And we need to pay for them without debt. And we need to get started ten years ago. The time bomb is ticking.

Show up to as many public hearings as you can, ask, ask, ask again, then ask again and again, yell, shout, push the issues, ask every official in your town, city, county, State and Federal Government. Run social media groups, bombard senators and members of congress, every day, every hour. Their phones should be lit up 24/7, their inboxes should be flooded. At the same time, find better candidates that want to save the planet and who understand basic UBI and GEC smart contracts.

ASK:

What is your plan to reduce our green house gas emissions? What is your plan to reduce our toxic pollution? What is your plan to clean up our lakes? What is your plan to clean up our beaches? What is your plan to build offshore windfarms? What is your plan to remove toxins from our rivers? What is your plan to remove plastic from our oceans? What is your plan to bring renewable energy to our State? What is your plan for more solar power in our county? When are you going to stop polluters? When will you change our school's curriculum to teach sustainable energy? What is your plan to stop global warming?

When will you stop killing us all?

Yeah, ask, ask, then ask again, push, and you will see the blank stares of stupidity.

Seriously, why not ask that last question? **WHEN will you stop killing us**!? Ask it out loud. Watch their faces. See the utter befuddlement and their immediate distraught reaction to try and flee, they will act as if you are the one that is crazy for asking that question.

If the people that we have running our schools, cities, towns, counties, agencies etc. if the people that we placed in charge of our future cannot answer the questions I posed here, then, well, then you have to admit that those people are not qualified to hold the jobs we gave them. Fire them. Replace them, now. What are you waiting for? We can impeach judges, even supreme court judges, we can recall governors, we can do a lot.

If anyone wants to run for office, any office, and they can't answer most of these questions, then they are not qualified. If you don't have a plan, you cannot be elected. **No plan, no ballot, no vote, no way**. If your whole point of being in office is to keep things the way they are, then you're killing us. And if you want to change things, tell us where you will get the money to pay for those changes. And then tell us why we should borrow that money.

If you don't have a plan for immediate and massive changes, then you don't deserve to be there, and you cannot be elected. If you can't grasp these simple concepts in this book, and you don't have a plan, then you are an obstructionist, you're committing a crime against humanity on a global level, and you are a co-conspirator with the toxin terrorists in your pocket and you cannot be elected. And, again, they will come with these responses:

"We cannot afford it; we don't have the money for all of that...."

"Where does money come from?"

"We create it."

"So why not create more so we can pay for all of this?"

"Because if we create too much money, then we have to pay it back, and we won't be able to afford the interest on it...."

How many must die due to Global Warming, pollution and toxin terrorists before you start asking questions. Just tell us, tell yourself, tell everybody you know:

"Guys, if we have another 100.000 people die because of global warming, then, I think then I will take a look at the issues."

"What you mean, 100.000?"

"Yeah, I think that is a fair number, don't you agree?"

"Nah, I think I get involved when we have at least 2million that died. But it has to be verifiable deaths due to climate change, or I won't believe it and I won't get involved."

"For me, I think I need to see more starving kids, you know, little children begging for food, you can see their despair, their misery, the hopelessness in their eyes and their parents are holding them and begging.... yeah, they should have had a better government, you know, like ours, we are thriving, we are making it, why can't they? I tell you what, I get involved when they tell me to get involved. Until then, let the government handle it, they know what to do."

"Yeah, this is all bogus man, show me some CEO who grabs some little girl, then shoves a handful of toxic mud in her mouth to watch her choke to death, just so he can make another million dollars, no way, companies don't do that... do they?"

"Ah, bullshit, get off it, all that rubbish talk about a warming planet, it's always been getting warmer, then it gets colder again, it will be alright, we just gotta get through this, why should we change anything for that, that makes no sense to me....I just want to keep going as it is and put more money in my account."

"We can't tell a company how to run their business. If their emissions are legal, why should we tell them suddenly they have to change everything?"

"You're just making a big fuss about nothing, we have plenty of space on our planet, it's huge, there are plenty of fish in the ocean, ice sheets are everywhere in the winter, it's nothing, you are just making this stuff up...."

If you are thinking or talking like that, you need to get off our planet. Get. Out. Now.

You know what else a UBI can do? This is the *main* reason most politicians don't like the idea of a universal basic income. <u>A UBI removes the fear of voting</u>. Read that again. It removes the *fear* of voting. If I don't vote for that schmuck, then some environmental talking head could shut down the factory where I work, then I will lose my income, my house, etc... I am so afraid, I am afraid to vote for what is best for me and my family and my children, etc... it is all fear based. The UBI removes that fear. It was fear that made people vote for Trump. Fear leads to anger; anger leads to hate....

I don't have to worry about the new elected officials shutting down the factory where I work, because it puts out too much pollution, so I could lose my job, lose my income etc. I don't have to be afraid anymore that I might not be able to make rent, or take my kids to the doctor... all that can be covered by UBI. UBI could pay for your basic rent, food and health insurance.... but if it does all that, and it can do so without debt, then politicians can't use the fear-based arguments against you and your family anymore. Do you see it now?

That means you no longer have to vote for some lying schmuck without a real plan.

What politician could possibly be against helping the very people in their own country being able to afford their basic needs, regardless of if they work or have a job. We have too many people on our planet already, there just aren't enough clean jobs for everyone. Why would you not want to help your own people, why would you not want to stop pollution?

But you are afraid to vote for what is best for you and all of us. You are. It's OK, you can admit it. Our official leaders are too corrupt. Maybe nothing will ever change. If that is the case, you might as well take your kids to Washington D.C., yeah, have a family vacation, go see the sites, be historic, watch where the power is being yielded. See a nice parade of politicians driving by and then ......

Joyce was the personal driver for Congressman Jones from Nebraska. He was elected 2years ago and still wasn't used to the way the media covered his ideas on global warming and how to gradually reduce more pollution.

He had sponsored a bill with some colleagues trying to slow down the reduction of toxic pollution allowance levels. The news organizations weren't too kind to him, he thought, as he was looking out at the all the people on the sidewalks, marching towards the capitol building when Joyce made a right turn.

"Wow, that's a lot of people here today... "he said.

"Yeah, but traffic is really light, looks like none of them came by car."

"By the way, how's the hand?"

"Much better, Sir. Thank you for asking."

"You hit that guy pretty hard, I mean thank you, I didn't want to get all those eggs on my suit, thank you, but damn, that sounded like you broke a brick in half..."

"Ha-ha, yes, Sir, it was a nice hit he received. Just doing my job, Sir."

The crowds on the sidewalk were getting bigger, it looked pretty packed. They were all chanting something and holding up signs, he could not make out what they said. He saw them staring into his SUV, trying to find out who was in it.

"Joyce, can you go a bit faster, I don't want them looking in the car like that...."

"No problem, Sir!" Joyce loved going faster, she had her foot down on the gas before the congressman finished his sentence. He could hear the engine revving up and for a moment he was pulled back into his seat...." wow" ...

Just for a second, she looked back at him, "are you OK Sir?"

"Ha-ha, yes, yes I am.... WATCH OUT!!!"

He could just see a small kid being flung in front of his car, just before they both felt the impact and the bouncing of the 2ton vehicle over the head of the kid that was laying there. The crowd outside screamed, Joyce screamed, and Congressman Jones said: "Oh, no, not again, oh my god no, why, why again......?"

Some people will figure that if our politicians are killing our children, they might as well do it in person.

How much carnage do you need to see or experience directly, before you get involved?

Change the system – change the future – save us - OR WE ALL DIE.

We all know that our ideas are crazy, but are they crazy enough? (Kobi Yamada)

The world has always gone forward when people have dared to have crazy ideas. (Gioconda Belli)

Good ideas are always crazy, until they're not. (Elon Musk)

Before you read the last chapter, I like to ask you to go to Amazon, log in, and leave a review for this book. I really like to know what you think about it. Am I too crazy for you, or not crazy enough? Please, be honest. Did you learn anything? Did you see things you never knew before?

One last chapter with more facts......

CHAPTER 17

The more you look into this, the more you get to confirm that we don't have much time left. I am sorry to inform you, here is the last report I will put in this book. This was December 8, 2020, I just read it on climate.gov, then I saw an article from Drew Kann, CNN.

Today's Arctic is much hotter, greener and less icy than it was even just 15 years ago, when NOAA published its first Arctic Report Card.

And with near-record high surface temperatures and near-record low sea ice observed yet again, the report card paints a picture of a region that is warming rapidly, at a pace far outpacing scientists' expectations.

"We thought the changes would take a lot longer, and the models were saying they would," said James Overland, an oceanographer at NOAA's Pacific Marine Environmental Laboratory, who has been a part of all 15 Arctic Report Cards and co-authored the portion on surface air temperatures in this edition. "But the rate of change we've seen in the last 20 years -- and especially the last five years -- is beyond what we thought would happen."

Last year saw another near-record-low sea ice extent, another sign that this air conditioner is breaking down, scientists say.

Sea ice freezes in winter and melts during summer, and this year's summer minimum extent was the second-lowest ever observed in the 42-year satellite record, according to the report. The trend of declines in the sea ice's winter maximum extent also continued this year, with March 2020's extent coming in as the 11th-lowest on record.

The 14 years from 2007 to 2020 have all seen the 14 lowest extents on record, and sea ice extents have declined by about 13% per decade since 1979.

"This isn't just like a low sea ice year or the permafrost thawing in on one place where the temperatures are rising -- **the *entire* ecosystem is changing**," Meier said. "And that's telling you that this isn't a fluke. It's something fundamental that's changing in the Arctic environment."

Would you be surprised to learn that China is building around 10 to 15 coal fired powerplants – in Africa? Yes, yes, they are. They are helping to finance and build them, to bring electricity to African countries, so their economy will grow, so they can then buy Chinese products, later. China is becoming their best friend, because some nations in Africa do not have the money to boost their economies. China is going to fill that void. Instead of American leadership with ideas on sustainable and renewable energy, you know, maybe solar panels, in the sun of Africa...... but no, why would politicians from the United States understand the need for energy in Africa, or that those are the emerging markets of our planet, oh good grief.

How much time do you think we really have left?

Here are some movies and Documentaries that you should watch, to learn more, or just for fun:
The Age of Consequences
Sand Wars
Freightened
Endgame2050
Pretty Slick, the gulf oil spill
Plastic, the real sea monster
Blue Gold, world water wars
Trouble the Water
This Changes Everything
Cows, Cash & Coverups
Losing Ground
Don't Flush your Freedom
Waste is Food
White Water Black Gold
Lobster War
My Name is Salt
Oil History Films
The Damned
Smoke & Fumes, the climate change cover up
Cape Spin
Tipping Point, the end of oil
Wood Industry, a busines against nature
Thorium 2011
Rising Tides
Sustainable
It Runs on Water
Thorium, the far side of nuclear power
Written on Water, the modern tale of a dry West
The Man who stopped the Desert
Circle of Poison
Flint, the poisoning of an American city

Tar Creek

If you watch one movie every day, within a month you will have more knowledge about our global issues than 99% of the people you know. Just one month! You just have to sit there and watch it. That's the easiest education you ever got.

Watch something about money too, highly entertaining,
The Big Short
Margin Call
Inside Job
Too Big to Fail
The Ascent of Money
Inside the Meltdown
Breaking the bank
The True Cost
Enron

You get the idea, after watching these, that our financial system is flawed as all get out and we are just pushing ourselves into more and more debt for no other reason than we created this system, and we are too lazy to fix it. Because there is money involved, and those that have it can give it to our politicians and stop them from ever fixing our system, it probably won't change any time soon. That's really scary, isn't it?

On a side note, for new projects that can help us all to live on this little planet a bit longer, we need better plants with more leaf surface areas, so they can absorb more CO_2, while also growing faster, with less water...   yeah, we can do that.   Anyone got $20mil lying around to get this started? I have some ideas; I've done the research.... Contact me if you can fund that, and yes, I will take GEC money, with pleasure.

Thank you for reading. Thank you so much.

www.ingramcontent.com/pod-product-compliance
Lightning Source LLC
Chambersburg PA
CBHW050247220526
45465CB00002B/578